T0135538

Face processing in congenital prosopagnosia

Dissertation

zur Erlangung des Grades eines
Doktors der Naturwissenschaften

der Mathematisch-Naturwissenschaftlichen Fakultät
und
der Medizinischen Fakultät
der Eberhard-Karls-Universität Tübingen

vorgelegt
von

Janina Esins
aus Schwerin, Deutschland

February – 2015

Bibliografische Information der Deutschen Nationalbibliothek

Die Deutsche Nationalbibliothek verzeichnet diese Publikation in der
Deutschen Nationalbibliografie; detaillierte bibliografische Daten sind
im Internet über http://dnb.d-nb.de abrufbar.

ISBN 978-3-8325-3983-2

Logos Verlag Berlin GmbH
Comeniushof, Gubener Str. 47,
10243 Berlin
Tel.: +49 (0)30 42 85 10 90
Fax: +49 (0)30 42 85 10 92
INTERNET: http://www.logos-verlag.de

Tag der mündlichen Prüfung: 23. April 2015.

Dekan der Math.-Nat. Fakultät: Prof. Dr. W. Rosenstiel
Dekan der Medizinischen Fakultät: Prof. Dr. I. B. Autenrieth

1. Berichterstatter: Prof. Dr. Heinrich H. Bülthoff
2. Berichterstatter: Dr. Andreas Bartels
Prüfungskommission: Prof. Dr. Martin A. Giese

                                Dr. Andreas Bartels

                                Prof. Dr. Christóbal Curio

                                Prof. Dr. Nikolaus F. Troje

I hereby declare that I have produced the work entitled: "Face processing in prosopagnosia", submitted for the award of a doctorate, on my own (without external help), have used only the sources and aids indicated and have marked passages included from other works, whether verbatim or in content, as such. I swear upon oath that these statements are true and that I have not concealed anything. I am aware that making a false declaration under oath is punishable by a term of imprisonment of up to three years or by a fine.

Janina Esins

Acknowledgements

First and foremost, I thank Dr. Isabelle Bülthoff and Dr. Johannes Schultz for their great supervision. Their constant support and encouragement helped me through some difficult phases of this thesis. And I am grateful for their patience in endless rounds of revisions during which they had to bear and improve my intellectual effusions.

I also thank Dr. Heinrich H. Bülthoff for giving me the opportunity to work at outstanding research facilities, the Max Planck Institute for Biological Cybernetics and the Department of Brain and Cognitive Engineering at the Korea University in Seoul. I very much enjoyed the great time at these amicable environments.

I would also like to acknowledge Prof. Martin Giese, and Dr. Andreas Bartels, as members of my advisory board, for comments and fruitful discussions of my work.

I owe special thanks to several other people helping to conduct my studies: Christian Wallraven for very interesting and fruitful discussions; Stephan de la Rosa for answering all my statistic questions; Karin Bierig for helping to conduct some experiments; Walter Heinz, Mirko Thiesen and Timo Hertel for making diverse computer problems easier to endure; Jacqueline Matzkeit and Katrin Prax for their incredibly efficient processing of necessary organizational duties; Joachim Tesch, Nina Gaißert, and Christoph Dahl for helping me out with stimuli; and Nele Hellbernd, Nack Duangkamol Srismith, BoRa Kim and Lea Ottenberger for helping me to create some other stimuli.

And of course I thank the whole MPI crowd for making working at the MPI so much fun with movie nights and holiday trips, coffee and tea breaks, fruitful scientific and private discussions, and several very interesting conversations over lunch ;) Thanks!

Most of all, I would like to express my deepest gratitude to my family, for always being there for me and supporting me in every way. You are always the first to call when something terrible happened; or something awesome. I am grateful for everything you have done for me. I love you.

Summary

Face recognition is one of the most important abilities for everyday social interactions. Congenital prosopagnosia, also referred to as "face blindness", describes the innate, lifelong impairment to recognize other people by their face. About 2 % of the population is affected, which means that one in fifty people shows noticeable problems in face recognition.

This thesis aimed to investigate different aspects of face processing in prosopagnosia in order to gain a clearer picture and a better understanding of this heterogeneous impairment. In a first study, various aspects of face recognition and perception were investigated to allow for a better understanding of the nature of prosopagnosia. The results replicated previous findings and helped to resolve discrepancies between former studies. In addition, it was found that prosopagnosics show an irregular response behavior in tests for holistic face recognition. We propose that prosopagnosics either switch between strategies or respond randomly when performing these tests. In a second study, the general face recognition deficit observed in prosopagnosia was compared to face recognition deficits occurring when dealing with other-race faces. Most humans find it hard to recognize faces of an unfamiliar race, a phenomenon called the "other-race effect". The study served to investigate if there is a possible common mechanism underlying prosopagnosia and the other-race effect, as both are characterized by problems in recognizing faces. The results allowed to reject this hypothesis, and yielded new insights about similarities and dissimilarities between prosopagnosia and the other-race effect. In the last study, a possible treatment of prosopagnosia was investigated. This was based on a single case in which a prosopagnosic reported a sudden improvement of her face recognition abilities after she started a special diet.

The different studies cover diverse aspects of prosopagnosia: the nature of prosopagnosia and measurement of its characteristics, comparison to other face recognition impairments, and treatment options. The results serve to broaden the knowledge about prosopagnosia and to gain a more detailed picture of this impairment.

Contents

I. Synopsis

Our faces tell who we are. When seeing a picture of a whole person without the face, the picture seems incomplete, and the person is unrecognizable in most cases. In contrast, a portrait showing just a face is perceived as a representation of the whole person. Our face carries our identity. It therefore bears great social importance. We do not only recognize people by their face, we also gain information about their age, gender, mood, and even judge their attractiveness and trustworthiness. Furthermore, faces contain a unique amount of social signals, like facial expressions, eye gaze direction, or attentional focus. These social signs are the most important cues that we use for everyday interaction with others.

The performance and reliability of our face recognition system is unrivaled by our other recognition systems, for example for objects. It is robust in correctly identifying a face in different lighting conditions, after years of ageing, or weight changes. At the same time, it is sensitive enough to distinguish within a split second between the thousands of faces of acquaintances and celebrities we know . Our face recognition system is so reliable that we only notice how important it is when it fails. The condition of a general impairment of face recognition is called 'prosopagnosia'.

This thesis examines various aspects of prosopagnosia, more precisely congenital prosopagnosia, the inborn form of the face recognition impairment. In three studies we investigated in what way congenital prosopagnosia impairs face processing, if it relates to other face recognition disturbances, and if it might be treatable.

This synopsis will first provide some background information on face processing in general and potential disturbances of it. Then information about prosopagnosia will be presented: its different manifestations, neurological causes, and possible treatments. Thirdly, the overall scope of this thesis and the findings of its three studies will be discussed in relation to previous work. The synopsis will be followed by three papers describing the methods, results and findings of the three studies in more detail.

1.   Face recognition

Faces are a very homogeneous object category, with eyes, nose, and mouth (features) arranged in a very similar way in all humans. This homogeneity of faces asks for specialized means in order to discriminate them from each other, e.g. to detect the subtle changes in form of the facial features and their spatial arrangement (configuration). A hypothesis about how humans achieve this outstanding performance in face recognition is that faces are processed holistically (Farah et al., 1998).

1.1.   Holistic processing

Holistic processing of faces means that the different components of a face (e.g., features and their configuration) are merged into a whole. It is very difficult to process single parts of a face individually without integrating other facial information (Maurer et al., 2002), and there is less part decomposition compared to object recognition (Tanaka and Farah, 1993; Lobmaier et al., 2010). One well-known demonstration of holistic face processing is the composite face illusion (Young et al., 1987). The top face half of one person combined with the bottom face half of another person gives the impression of a new, third identity. It is very hard to process the two different parts individually to identify the original persons (Figure 1), even if they are well-known to the viewer. Misaligning the two face halves makes the illusion disappear.

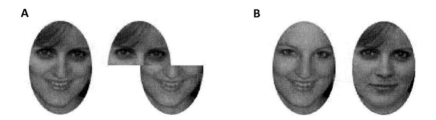

Figure 1: (A) Composite illusion. Aligning the top and bottom face half of two individuals creates the illusion of a new, third person. Misaligning the two halves makes the illusion disappear. (B) Original faces used for the illusion in panel A.

As holistic processing depends on the integration of the different dimensions of facial information (e.g. features, configuration, etc.), the disturbance in retrieval of information from any of these dimensions impairs holistic processing and thus face recognition. This has been shown for example in studies using blurred and scrambled face stimuli, in which either featural or configural information are respectively disrupted (Collishaw and Hole, 2000).

1.2. The other-race effect

One example of deficits in face recognition, known by nearly every human, can be experienced when looking at other-race faces. Faces of a foreign ethnicity are harder to distinguish from each other than faces of the own, familiar ethnicity (Malpass and Kravitz, 1969). This phenomenon is called the 'other-race effect', 'own-race-bias', or 'cross-race effect'. One possible and widely accepted explanation for this effect is the higher level of expertise for same-race faces compared with other-race faces (Meissner and Brigham, 2001).

Studies have shown, that other-race faces are processed less holistically than own-race faces (Michel et al., 2006b), and that there is an own-race advantage for both configural and featural processing (Hayward et al., 2008). These findings support the hypothesis that face recognition is compromised if holistic processing or processing of some of the facial information (e.g. features, configuration, etc.) is impaired.

Congenital prosopagnosia, a more severe impairment of face recognition, is the research focus of this thesis and will be discussed in the next section.

2. Congenital prosopagnosia

A general impairment of face recognition is summarized under the term 'prosopagnosia'. The term 'prosopagnosia' comes from the two Greek words 'prosopo' (πρόσωπο) which means 'face' and 'agnosia' (αγνωσία) which means 'not knowing'. Prosopagnosia is also often referred to as 'face blindness', though this term can be misleading. Even though prosopagnosics cannot read the identity coded by a face, or only with great effort, they are able to see and detect faces (Garrido et al., 2008).

2.1. Forms and occurrence frequency of prosopagnosia

Prosopagnosia can be acquired, caused by an acute brain lesion damaging a previously functioning face recognition system. However, this thesis investigates the congenital, innate form of prosopagnosia: in the people affected, the face recognition system never developed or worked normally. Approximately 2 % of the population is affected by this form of prosopagnosia (Kennerknecht et al., 2006; Kennerknecht et al., 2008a). Several aspects of the impairment indicate that congenital prosopagnosia might be hereditary (Grüter, 2004), as many investigated cases have first-order relatives, who also exhibit face recognition deficits (e.g. Kennerknecht et al., 2008b; Schmalzl et al., 2008a). Another term to refer to prosopagnosia with an early onset is 'developmental' prosopagnosia. This term is often used interchangeably with 'congenital' prosopagnosia (Duchaine et al., 2007a); however, it is occasionally used to refer to instances where prosopagnosia was likely acquired in early childhood, due to brain damage (Barton et al., 2003). For this reason, in this thesis the term 'congenital' is used to underline the innate, lesion-free form of prosopagnosia for the participants of the studies. Participants' self-testimony or parents' testimony was used to determine the lack of brain lesions. Additionally, in many cases participants reported first-order relatives with face recognition impairments, strengthening the classification as hereditary, congenital form of prosopagnosia. In the further course of this thesis, 'prosopagnosia' will refer to the congenital from, if not stated otherwise.

2.2. Manifestations of prosopagnosia

The degree of impairment severity differs between prosopagnosics. Very likely, face recognition ability is Gaussian distributed among the population with prosopagnosics at the low end of the spectrum (Russell et al., 2009; Kennerknecht et al., 2011). At the other end of the spectrum are so-called super recognizers, who are able to recognize a face after decades, even if they have just very briefly met that person (Russell et al., 2009) (Figure 2).

In most studies, prosopagnosia is diagnosed if the performance in face recognition tests is two standard deviations below the mean performance of a control population. No standard tests exist for the diagnosis of prosopagnosia; however, it became common practice to use the Cambridge Face Memory Test (CFMT, Duchaine and Nakayama (2006)), mostly in combination with further face recognition tests, like tests of famous face recognition and face perception (Rivolta et al., 2011; Tree and Wilkie, 2010; Shah et al., 2015). In addition, other means of diagnosis exist, for example questionnaires (Grüter, 2004; Stollhoff et al., 2011).

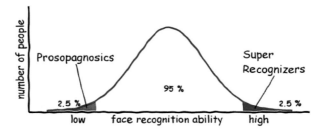

Figure 2: Graphical representation of the possible normal distribution of face recognition ability. Performance outliers are prosopagnosics (low end of the scale) and super recognizers (high end of the scale).

Prosopagnosia includes all cases with generally impaired face recognition, no matter what the underlying cause is. A study found a prevalence of subjectively perceived face recognition difficulties in 47 % of participants with Asperger's syndrome compared to 11 % of controls (Nieminen-von Wendt et al., 2005). Another study investigating the comorbidity of prosopagnosia in patients with social developmental disorders found that some, but not all participants had impaired face recognition compared to controls (Barton et al., 2004). Some patients with schizophrenia were found to have impaired face recognition (Archer et al., 1992), as well as adults with attention deficit hyperactive disorder (ADHD) (Markovska-Simoska and Pop-Jordanova, 2010).

Several studies have investigated the nature of impairments occurring in prosopagnosia and what impact prosopagnosia has on different aspects of face recognition. As with the other-race effect, it has been shown that prosopagnosia impairs holistic face processing in general (Avidan et al., 2011), as well as configural and featural processing in particular (Hayward et al., 2008). However, the results were mixed, with prosopagnosics showing large variations in impairments. In addition, when assessed with several tests, prosopagnosics showed no systematic pattern of impairment: They were impaired in some tests and performed normally in others, without a clear structure at the individual and group level, and without relation to their face recognition in the 'real world' (Le

Grand et al., 2006; Schweich and Bruyer, 1993). The pattern of impairments even varies across prosopagnosics belonging to the same family, for which one would expect similar traits and impairment patterns (Schmalzl et al., 2008a). Investigated aspects with non-uniform findings include holistic processing (e.g. assessed by the composite face test), processing of features and configuration, recognition of facial expression, recognition of gender, judgment of attractiveness, viewpoint matching, etc. All cases of prosopagnosia have in common that their face recognition is impaired, but more specific impairments strongly vary from case to case.

Due to their difficulties in recognizing faces, prosopagnosics develop different recognition strategies based on means like distinctive facial features (e.g. aquiline nose, thick eyebrows), voice, hairdo, gait or obvious facial blemishes to recognize someone (Rodrigues et al., 2008; Mayer and Rossion, 2009). Of course, these strategies for face recognition are also used in face recognition tests (Stollhoff, 2010; Duchaine et al., 2003). By applying their compensatory strategies, some prosopagnosics have been reported to obtain 'normal' scores in the CFMT (Bate et al., 2013) and in famous face recognition tests (Rodrigues et al., 2008). These strategies could potentially falsify tests results, being one source of the heterogeneity of the impairment. To the best of our knowledge, this problem has not yet been investigated. However, strategies cannot explain all findings of heterogeneity: a study testing very young, related prosopagnosics (4, 5, and 8 years of age) also found heterogeneous impairment patterns (Schmalzl et al., 2008a). The authors concluded that the heterogeneity is genetically intrinsic to prosopagnosia, rather than a result of strategies developed during life, as the participants were too young to have developed consistent strategies.

2.3. Neurophysiological and genetic basis of congenital prosopagnosia

Face recognition involves several interconnected brain regions, the so-called 'face processing network'. This network has three core areas in each hemisphere (see Figure 3). The fusiform face area (FFA) is located in the lateral fusiform gyrus of the temporal lobe and is believed to process facial identity (Kanwisher and Yovel, 2006). Another core face area is the occipital face area (OFA) which is located in the inferior occipital gyrus. The OFA has been linked to the early visual processing of faces (Pitcher et al., 2011) and to providing input to the FFA (Haxby et al., 2000). The third core area for face recognition is the superior temporal sulcus (STS) which processes biological and facial motion and gaze direction (Allison et al., 2000; Hoffman and Haxby, 2000). This core network is extended by further face areas processing person knowledge, emotional aspects, etc. (Ishai et al., 2005).

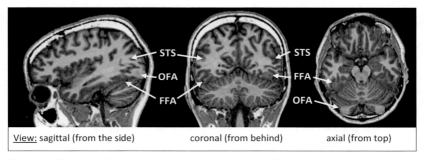

| View: sagittal (from the side) | coronal (from behind) | axial (from top) |

Figure 3: Overview of the core areas of the face network and their approximate position in the authors brain.

Brain activity measurement techniques like electro-encephalography (EEG) or magnetic resonance imaging (MRI) help to shed light onto the underlying neurophysiological causes of prosopagnosia. Several studies investigating prosopagnosics found reduced grey matter volume in several regions or in the whole of the temporal lobe (Bentin et al., 1999; Behrmann et al., 2007; Garrido et al., 2009). Furthermore, structural connections between STS and other core face areas were reported to be reduced in prosopagnosics (Thomas et al., 2009; Pyles et al., 2013), as was the diffusivity of the tracts connecting the FFA with the other core face areas (Gomez et al., 2015). Whether the reduced volume of, and connectivity between, the core faces areas in prosopagnosia are causes or consequences of prosopagnosia remains an open question.

The occurrence pattern of prosopagnosia within families was first interpreted as indication for an autosomal dominant inheritance process (i.e. if one parent is affected, chances are 50 % for each child to be affected as well) (Grüter et al., 2007). However, potential candidate genes remain to be identified. Finding the candidate genes and their role in developing or maintaining face recognition will be a dramatic breakthrough in the understanding of cognition and may even result in treatment opportunities for prosopagnosia.

2.4. Treatments of prosopagnosia

Several attempts have been made to improve face recognition in the acquired as well as the congenital forms of prosopagnosia. Two different approaches have been undertaken so far to treat congenital prosopagnosia: (1) training to improve face recognition, and (2) some form of 'medication'.

(1) In some cases, the training was aimed at helping to identify unique facial characteristics useful to learn and recognize faces (Brunsdon et al., 2006; Schmalzl et al., 2008b). In both studies, training only brought improvement for the trained faces and a generalization to untrained faces was not possible. In other cases, training was aimed at helping prosopagnosics to improve their holistic face processing by extracting spatial information (DeGutis et al., 2014). The results were mixed with some of the prosopagnosics being able to raise their performance level to that of controls, while some prosopagnosics were not able to improve their performance at all.

(2) So far there is only one study that has reported treating prosopagnosics using some form of 'medication': the hormone oxytocin (Bate et al., 2013). Oxytocin has been found to be crucial for various social behaviors (Lee et al., 2009) and seems to enhance memory for faces (Savaskan et al., 2008). In the study by Bate and colleagues (2013) some prosopagnosics significantly, temporarily improved their face recognition performance while others showed only very little to no improvement.

3. Thesis overview and discussion

3.1. Motivation

The described heterogeneity of prosopagnosia in its manifestations, and its response to treatments gave rise to the topic of this thesis. Due to the conflicting results reported in previous studies and the heterogeneous pattern of impairments, the possible existence of subgroups in prosopagnosia was proposed (e.g. Avidan et al., 2011). However, why subgroups might exist had never been specified. The discovery of the genes responsible for the different subgroups would be a quantum leap in the research field. Unfortunately, the search for possible candidate genes might prove more difficult than anticipated, if, for example, prosopagnosia is a symptom resulting from various causes. For this reason, a categorization into subgroups might help to pre-sort the individual prosopagnosic cases, before searching for common genetic factors within the subgroups. Therefore, the goal of this thesis was to create an extended test battery to investigate the heterogeneous patterns of impairments in prosopagnosia in more detail, to broaden and expand the understanding of this disorder, and to investigate the possibility to detect at least some subgroups based on their pattern of impairments in psychophysical tests.

3.2. Test battery

To achieve the goals of this thesis, an extended battery containing 16 tests was designed. The test battery consists of a mix of face and object recognition test, some well-established and some newly created, as well as some tests known from previous studies to yield non-uniform results for prosopagnosic participants. Especially the latter category of tests was considered helpful to detect different forms of prosopagnosia, thus facilitating the search for subgroups. The battery was used to test a comparatively high number of prosopagnosics[1] as well as age-, gender and education matched controls and other-race observers.

The test battery consists of 16 tests (references are given for tests and stimuli graciously provided by other researchers):

1. Cambridge Face Memory Test (CFMT), a test of face memory and holistic processing (Duchaine and Nakayama, 2006b)
2. Cambridge Car Memory Test (CCMT), a test of object memory and processing (Dennett et al., 2011)
3. Surprise recognition test, a test of holistic processing and the unconscious intake of facial identity information
4. Composite face test, a test of holistic processing
5. Similarity rating test, a test of the sensitivity to featural and configural facial information
6. Face gender test, a test of gender recognition on the basis of the face

---

[1] The prosopagnosic participants were diagnosed by the Institut für Humangenetik, Westfälische-Wilhelms-Universität, Münster, Germany, based on a screening questionnaire and a diagnostic semi-structured interview (Stollhoff et al., 2011).

7. Facial motion advantage test, a test of holistic processing and the advantage of dynamic information on face recognition (O'Toole et al., 2005)

8. Object and face test, a test of object and face recognition

9. Other-race test, a test of holistic processing and recognition of own- and other-race faces (Michel et al., 2006a)

10. Covert face recognition, a test of holistic processing and implicit, unconscious face recognition

11. Face categorization test, a test of holistic versus feature-based processing of faces (Schwarzer et al., 2005)

12. Facial idiosyncrasy test, a test of face recognition based purely on facial motion (Dobs et al., 2015)

13. Navigation test, a test of the sense of orientation

14. Long term memory test, a test of the long term memory (i.e. 2 years) of faces and objects

15. Facial imagery, a test of the mental imagination ability for faces

16. Famous face test, a test of holistic processing and recognition of familiar faces

This test battery was used to conduct two studies. One study investigated the nature of the heterogeneous manifestations of prosopagnosia and if it is possible to detect potential subgroups based on the performance in those tests. This study is described in detail in chapter II. Another study compared two phenomena of impaired face recognition performance: prosopagnosia and the other-race effect. This study is described in detail in chapter III.

A last study investigated the occurrence of improvement of face recognition abilities in a single prosopagnosic case. This study is described in chapter IV. All three studies are summarized in the following sections, followed by two sections describing the search for subgroups and generally discussing the results of the research conducted for this thesis.

3.3. Face perception and test reliability in prosopagnosia

In the study described in chapter II, we investigated different aspects of face perception with tests 1 to 7 (listed on page 9) for prosopagnosic and control participants. Besides analyzing the test performance, we also compared the reliability of the different tests across participant groups.

As a group, prosopagnosics showed a number of perceptual impairments compared to controls: impaired holistic processing impaired processing of featural and configural information of faces, impaired performance in gender recognition, and a reduced advantage for recognition of faces in motion. There was no difference between groups in object recognition and in the unconscious intake of face identity information. These results mostly confirmed our expectations and some tests replicated or confirmed findings of previous studies.

For each test, we calculated its internal consistency reliability, which is an indication of the test's quality: the trials of a test measuring the same face recognition mechanisms should produce similar scores. We were surprised to find different reliability coefficients between groups: prosopagnosics showed much lower reliability coefficients than controls in tests of holistic face processing. This suggests that the prosopagnosic participants had a different response behavior quality, exhibiting a very irregular and changing response behavior within these tests. A possible explanation is that prosopagnosics switch between strategies or respond randomly.

This finding has implications for the interpretation of prosopagnosics' test results. First, individual test scores are used in the literature to compare prosopagnosics with controls; second, they are used to compare prosopagnosics with each other; and third, they are used to analyze prosopagnosics' response patterns in order to characterize their impairment and search for subgroups. We believe that the cause for this reduced reliability needs to be investigated and more reliable tests need to be designed, to provide robust and less noisy test results, to reliably investigate prosopagnosia and its impairments.

3.4. Comparing the other-race effect and prosopagnosia

Face recognition is a robust ability, yet it can be affected by different causes. Prosopagnosia and the other-race effect are both characterized by impaired face recognition, and share apparent commonalities. For both groups, holistic processing is reduced compared to controls (Avidan et al., 2011; Rhodes et al., 1989; Michel et al., 2006b), as are featural and configural processing (Lobmaier et al., 2010; Hayward et al.,

2008). Therefore, the other-race effect is sometimes used as an example to explain the impairments prosopagnosics experience in everyday life (e.g. Kennerknecht, 2011).

As prosopagnosia and looking at other-race faces both lead to reduced face recognition performance, the same face recognition mechanisms could be disturbed in both cases, although through different causes. To investigate this hypothesis, we compared whether prosopagnosia and the other-race effect impair featural and configural face processing in a similar way. To that end, we tested German prosopagnosics, Korean participants and German controls on their holistic, featural and configural processing of Caucasian face stimuli, as well as object recognition. In the study described in chapter III we report the performance of these groups on tests 1 (CFMT), 5 (Similarity rating), and 8 (Object and face test), as listed on page 9.

Based on our results, we could disprove the hypothesis that a common underlying mechanism is responsible for the impairments observed in prosopagnosia and the other-race effect. Korean and prosopagnosic participants exhibited different patterns in impairment of featural and configural processing of Caucasian faces. The sensitivity to the featural changes was about the same for both groups. The sensitivity to configural changes however was only impaired for prosopagnosics, while the Koreans were as good as the German controls. Further, we could show that prosopagnosia has a stronger impact on face processing than does observing other-race faces, as prosopagnosics were significantly more impaired in recognition of Caucasian faces than Koreans, who in turn were significantly more impaired than the German controls.

Additionally, we were able to gain new insight into general face recognition mechanisms. Our results suggest that sensitivity to features is not crucial for determining face recognition abilities. This is indicated by the fact that we found no difference between Koreans and prosopagnosics in their sensitivity to facial features, while prosopagnosics were more strongly impaired in general face recognition abilities than Koreans. However, we could substantiate the hypothesis that configural sensitivity relates to face recognition ability (Freire et al., 2000), because prosopagnosics were more strongly impaired in general face recognition and configural processing than Koreans.

Overall, this study provided new insights into the face processing disturbances caused by prosopagnosia and the other-race effect, and into general face processing mechanisms.

3.5. Galactose uncovers face recognition

So far, only one study reported that a form of medication (administration of oxytocin) was able to improve face recognition abilities in some but not all participating prosopagnosics (Bate et al., 2013). Therefore, the discovery by a participant of our studies, LI, that her face recognition improved due to a change in her diet, led to further investigations of that case. This case report is described in more detail in chapter IV.

LI had added about 5 gram of galactose to her daily nutrition. The reason for this was self-medication for her attention deficit hyperactivity disorder (ADHD), and was unrelated to her face recognition deficits.

Galactose is a simple sugar with natural occurrence, for example as a component of the milk sugar lactose. Healthy humans synthesize two to ten gram of galactose per day. It was found that oral intake of galactose prevents cognitive deficits in mice induced with Alzheimer's disease symptoms (Salkovic-Petrisic et al., 2014). Furthermore, descriptions of several cases of positive effects of the oral intake of galactose for patients with Alzheimer's or Parkinson's disease exist (Mosetter, 2008).

LI reported that through galactose not only her face recognition abilities improved, but that in addition, amongst other things, her mental imagery became more vivid and her sense of orientation increased. The effects vanished after stopping the intake of galactose. To investigate the universality of this treatment 16 additional congenital prosopagnosics were recruited to tests the possible impact of galactose on their condition. None of them reported any noticeable effects. The single occurrence of a galactose-treatable case of prosopagnosia might be another indicator of the existence of diverse subgroups.

However, we also completed the following (unpublished) study to investigate the objective effects of galactose in LI. We conducted a double blind placebo study in two sessions with existing and newly created tests of face recognition, navigation, and mental imagery (see Table 1). We provided LI with a set of seven 5-gram-doses of galactose and another set with doses of placebo (starch). To avoid identification based on looks or taste, both agents were packed in identical-looking medication capsules by a pharmacy in Münster. LI was asked to stop her private galactose intake for four weeks before starting the seven-day 'medication' with the first set of capsules. On the day after the seventh intake, LI was tested and interviewed in a first session by an experimenter naïve as to the treatment used. Then another four weeks without private galactose followed, after which the same procedure was repeated with the second set of capsules and the second testing session. For each test, two versions were used with different stimuli for both sessions. In pilot studies, the stimuli were adjusted so that the tests had a similar degree of difficulty in both sessions.

We found nearly no significant differences[2] between the results of tests performed after galactose or starch intake ($ps > .05$ for all but two tests; see Table 1 for details). Several explanations could account for this lack of an effect. Perhaps our tests were not adequate to measure the effects of galactose. Alternatively, galactose may have different effects on subjective and objective performance in face recognition. However, some aspects of the results hint towards a mishap in the test methodology (e.g. a mix-up of the capsules by the experimenters or the pharmacy): First, LI reported experiencing the known and expected effects of galactose in the placebo session (session 2), but not in the galactose session (session 1, see Table 1). Second, significantly better results were obtained in two of nine tests in the placebo session, with other tests showing a similar trend. The latter could be due to training effects, because of the repetition of session 1 tests in session 2. Alternatively, we cannot exclude the possibilities that the effects of galactose for LI vary depending on another factor that is yet to be discovered. Overall, the double-blind experiment unfortunately gave no experimental evidence that corroborates the subjective effects of galactose on face recognition experienced by LI. Therefore, our initial findings published as indicated in chapter IV (Esins et al., 2013) should be taken with caution. Further tests are necessary to verify if galactose has positive effects and can serve as a possible therapeutic method for prosopagnosia.

---

[2] For each test, we calculated the difference in performance between session 1 and 2 for LI. This difference was then compared to the performance differences achieved by the participants of the pilot studies on the respective test items. The Bayesian hypothesis statistical methods used for the statistical analysis are described in (Crawford et al., 2011) and were conducted with the provided computer programs downloaded from
http://homepages.abdn.ac.uk/j.crawford/pages/dept/SingleCaseMethodsComputerPrograms.HTM.

Table 1: Tests and results of the double-blind study to investigate the measurable effects of galactose

	Test	Session 1 GALACTOSE	Session 2 PLACEBO	p-value [2] (one-tailed)	Information about the test	Stimulus examples
A	Australian Face Memory Test (McKone et al., 2011)	78 % correct	81 % correct	$p = .40$	Faces of Australian Caucasians to memorize and recognize among distractors	
B	Chinese Face Memory Test (McKone et al., 2012)	56 % correct	69 % correct	$p = .062$	Faces of Chinese Asians to memorize and recognize among distractors	
C1	*High* imagery sentences (adapted from (Eddy and Glass, 1981))	88 % correct	88 % correct	$p = .13$	Multiple choice sentences which need mental imagery to be answered correctly.	A tractor has the large wheels in front / in the back / in front and back
C2	*Low* imagery sentences	75 % correct	100 % correct	$p = .005*$	Multiple choice sentences which need NO mental imagery to be answered correctly.	A week has five / six / seven days.
D1	Navigation - Number of wrong turns	7	7	$p = .27$	Remember and recall the correct way through a labyrinth (steering similar to an ego-shooter computer game). Walking time and number of incorrect turns are measured. Same as test number 5 listed on page 9.	
D2	Navigation - Mean walking time through the labyrinth	81.7 s	73.5 s	$p = .32$		
D3	Navigation - Scene recognition	0.14 d' score	1.08 d' score	$p = .086$	Recognition of street corners in a labyrinth after walking through it	
E1	Mental rotation – accuracy	98 % correct	98 % correct	$p = .21$	Mental rotation – recognition if rotated letters and symbols are mirror-inverted.	
E2	Mental rotation – reaction time	1.53 s	1.24 s	$p = .018*$	Mental rotation – mean recognition time	
F	Vividness of Visual Imagery (Marks, 1973)	4.85	1.10	-	Rating of strength of a mental image on a Likert-scale from 1 to 5 (low value = high self-reported imagery)	The sun is rising above the horizon into a hazy sky
G1	Sensitivity to facial *features*	0.86	-0.53 invalid[†]		Same as test number 5 listed on page 9 and described in detail in chapter III (Esins et al., 2014b)	
G2	Sensitivity to facial *configuration*	0.49	-0.23 invalid[†]			
H	Interview	No effects noticed	Strong effects noticed	-	Short summary of self-reported, perceived effects	What changes did you perceive since the intake of the capsules?

[2] Footnote, see previous page.

* Significant p-values are marked with an asterisk.

[†] Results of test G, session 2 are invalid, because the result values are negative. LI reported fatigue and reluctance to perform this test in the second session, as it is a rather long (45 min) and tiring test.

3.6. Search for subgroups

As prosopagnosia was stated to be a heterogeneous disorder (Le Grand et al., 2006; Schweich and Bruyer, 1993), we assembled the extended test battery to investigate the possibility to detect subgroups based on patterns of impairment in psychophysical tests. We also found very heterogeneous impairments amongst our prosopagnosics participants in the single tests, but we were not able to find signs of common patterns. We used different mathematical approaches including, amongst others, principal component analyses, various types of cluster analyses and correlation analyses (Esins et al., 2012). However, we were able to reveal that prosopagnosics' response behavior is very irregular as shown by the reduced reliability for prosopagnosics in tests of holistic face processing (section 3.3 and chapter II). This response behavior leads to noisy results, which might obstruct finding regularities in response patterns, on which the detection of subgroups is based. Therefore, we suggest that this irregular response behavior is a possible impediment to detect subgroups using current tests for holistic face recognition.

3.7. General discussion

This thesis investigated different aspects of prosopagnosia to broaden and advance the understanding of this disorder. We assembled an extended test battery and used it to examine the impairments caused by prosopagnosia in different face recognition tasks (chapter II), and to find differences in the underlying causing mechanisms for prosopagnosia and the other-race effect (chapter III). We also investigated a single case of prosopagnosia that appeared to be treatable by oral intake of galactose (chapter IV).

The studies also broadened our understanding of face recognition in general. In chapter III, we show that the face recognition impairments caused by prosopagnosia are different from the face recognition deficits occurring when looking at other-race faces. Prosopagnosia does not only have a stronger impact on face recognition in general, but also on configural processing of faces. More generally, our results implicate that sensitivity to facial features is not crucial for determining face recognition abilities, while sensitivity to facial configuration seems to be linked to face recognition abilities.

Furthermore, our studies are the first to report a possible positive effect of galactose on one case of prosopagnosia, as described in chapter IV. However, as we could not verify this effect experimentally, this finding needs further confirmation.

Another new finding, first reported by our study described in chapter II, is the reduced reliability for prosopagnosic participants in tests of holistic processing of static faces. Based on our results it is not possible to point out the actual cause for this reduced reliability. However, we suggest that a possible source is an irregular, changing response behavior of prosopagnosics within the tests, for example as a result of the usage of different compensatory strategies or random responses. The fact that prosopagnosics use strategies has been reported before (Stollhoff, 2010; Duchaine et al., 2003). Our study is the first to show that there are indeed qualitative differences in test responses for prosopagnosics and controls. This irregular response behavior of prosopagnosics indicates that their test results might not reflect their actual abilities but rather the customary usage of various strategies or random responding. This is an important finding, meaning that implications drawn upon these results by future and previous studies should be taken with caution. In the light of the finding of an irregular response behavior by prosopagnosics, we will now look back at our two other studies. For the study comparing the other-race effect with prosopagnosia (chapter III), we argue that our conclusions are nevertheless valid. We based our conclusions partly on the results of the similarity rating (testing featural and configural processing; test 5 listed on page 9 and described in detail in chapter III). This test was shown to have excellent reliability coefficients for prosopagnosics and controls (see chapter II, Table 4). For the Koreans of the study in chapter III, reliability coefficients of the similarity rating were > .94 for Cronbach and split-half estimates for featural and configural condition. We also based our conclusions partly on the results of the CFMT, which was found to have a significantly reduced reliability for prosopagnosics (chapter II). However, we only interpreted the results on a group-wise level, as we argue that, despite low reliability, the tests are suitable for a coarse comparison of face processing abilities between groups. Looking back at the galactose study (chapter IV), the finding of an irregular response behavior in prosopagnosics provides another possible explanation for the absence of an effect of galactose in LI's test results. A possible irregular response behavior of LI could have added noise to the results, thus reducing the chances to find an effect (irrespective of the possible mishap).

Apart from our studies, the finding of reduced test reliability for prosopagnosic participants also has several implications for general research on prosopagnosia. The irregular response behavior could be an explanation for the unsystematic pattern of impairments, which we observed in our tests (see section 3.6) and which has previously led others to the conclusion that prosopagnosia is a very heterogeneous disorder (Le Grand et al., 2006; Schweich and Bruyer, 1993; Schmalzl et al., 2008a). This heterogeneity was suggested to be intrinsic to prosopagnosia (Schmalzl et al., 2008a). Our findings suggest that the irregular response behaviors exhibited by most prosopagnosics could be another possible cause for the heterogeneity of prosopagnosia. Furthermore, this response irregularity leads to noisy test results. This in turn could obstruct the search for subgroups of prosopagnosia that relies on finding regularities in response patterns. We suggest that it might not be possible to detect subgroups of prosopagnosia with the tests for holistic face processing developed so far.

4. Outlook

Future research should explore the differences in test reliability for prosopagnosic and control participants and their causes in further detail. Better tests with higher reliabilities for prosopagnosics need to be designed. In addition, larger sample sizes will be needed for further investigations of prosopagnosia. Better tests and larger participant groups could help to increase the signal-to-noise ratio in test results needed for the detection of subgroups. If subgroups are identified and their respective cause becomes known, it might even be possible to find specific treatments for subgroups with treatable causes (e.g. nutritional deficits) to improve the life of the people affected.

References

Allison, T., Puce, A., and Mccarthy, G. (2000). Social perception from visual cues: role of the STS region. *Trends Cogn. Sci.* 4, 267–278. doi:10.1016/S1364-6613(00)01501-1.

Archer, J., Hay, D. C., and Young, A. W. (1992). Face processing in psychiatric conditions. *Br. J. Clin. Psychol.* 31, 45–61. doi:10.1111/j.2044-8260.1992.tb00967.x.

Ariel, R., and Sadeh, M. (1996). Congenital Visual Agnosia and Prosopagnosia in a Child: A Case Report. *Cortex* 32, 221–240. doi:10.1016/S0010-9452(96)80048-7.

Avidan, G., Tanzer, M., and Behrmann, M. (2011). Impaired holistic processing in congenital prosopagnosia. *Neuropsychologia* 49, 2541–2552. doi:10.1016/j.neuropsychologia.2011.05.002.

Avidan, G., Tanzer, M., Hadj-Bouziane, F., Liu, N., Ungerleider, L. G., and Behrmann, M. (2013). Selective Dissociation Between Core and Extended Regions of the Face Processing Network in Congenital Prosopagnosia. *Cereb. Cortex* 24, 1565–78. doi:10.1093/cercor/bht007.

Barton, J. J. S., Cherkasova, M. V, Hefter, R. L., Cox, T. A., O'Connor, M., and Manoach, D. S. (2004). Are patients with social developmental disorders prosopagnosic? Perceptual heterogeneity in the Asperger and socio-emotional processing disorders. *Brain* 127, 1706–16. doi:10.1093/brain/awh194.

Barton, J. J. S., Cherkasova, M. V, Press, D. Z., Intriligator, J. M., and O'Connor, M. (2003). Developmental prosopagnosia: A study of three patients. *Brain Cogn.* 51, 12–30. doi:10.1016/S0278-2626(02)00516-X.

Bate, S., Cook, S. J., Duchaine, B. C., Tree, J. J., Burns, E. J., and Hodgson, T. L. (2013). Intranasal Inhalation of Oxytocin Improves Face Processing in Developmental Prosopagnosia. *Cortex*, 1–9. doi:10.1016/j.cortex.2013.08.006.

Behrmann, M., and Avidan, G. (2005). Congenital prosopagnosia : face- blind from birth. *Trends Cogn. Sci.* 9, 180 – 187.

Behrmann, M., Avidan, G., Gao, F., and Black, S. (2007). Structural imaging reveals anatomical alterations in inferotemporal cortex in congenital prosopagnosia. *Cereb. Cortex* 17, 2354–63. doi:10.1093/cercor/bhl144.

Behrmann, M., Avidan, G., Marotta, J. J., and Kimchi, R. (2005). Detailed exploration of face-related processing in congenital prosopagnosia: 1. Behavioral findings. *J. Cogn. Neurosci.* 17, 1130–49. doi:10.1162/0898929054475154.

Bentin, S., Deouell, L. Y., and Soroker, N. (1999). Selective visual streaming in face recognition : evidence from developmental prosopagnosia. *Neuroreport* 10, 823–827.

Bodamer, J. (1947). Die Prosop-Agnosie. *Arch. für Psychiatr. und Nervenkrankheiten Ver. mit Zeitschrift für die Gesamte Neurol. und Psychiatr.* 179, 6–53. doi:10.1007/BF00352849.

Bonett, D. G. (2003). Sample Size Requirements for Comparing Two Alpha Coefficients. *Appl. Psychol. Meas.* 27, 72–74. doi:10.1177/0146621602239477.

Bowles, D. C., McKone, E., Dawel, A., Duchaine, B. C., Palermo, R., Schmalzl, L., Rivolta, D., Wilson, C. E., and Yovel, G. (2009). Diagnosing prosopagnosia: effects of ageing, sex, and participant-stimulus ethnic match on the Cambridge Face Memory Test and Cambridge Face Perception Test. *Cogn. Neuropsychol.* 26, 423–55. doi:10.1080/02643290903343149.

Brainard, D. H. (1997). The Psychophysics Toolbox. *Spat. Vis.* 10, 433–436. doi:10.1163/156856897X00357.

Brown, W. (1910). Some Experimental Results in the Correlation of Mental Abilities1. *Br. J. Psychol. 1904-1920* 3, 296–322.

Brunsdon, R., Coltheart, M., Nickels, L., and Joy, P. (2006). Developmental prosopagnosia: A case analysis and treatment study. *Cogn. Neuropsychol.* 23, 822–40. doi:10.1080/02643290500441841.

Chatterjee, G., and Nakayama, K. (2012). Normal facial age and gender perception in developmental prosopagnosia. *Cogn. Neuropsychol.* 29, 482–502. doi:10.1080/02643294.2012.756809.

Collishaw, S. M., and Hole, G. J. (2000). Featural and configurational processes in the recognition of faces of different familiarity. *Perception* 29, 893–909. doi:10.1068/p2949.

Crawford, J. R., Garthwaite, P. H., and Ryan, K. (2011). Comparing a single case to a control sample: testing for neuropsychological deficits and dissociations in the presence of covariates. *Cortex.* 47, 1166–78. doi:10.1016/j.cortex.2011.02.017.

Dalrymple, K. a., Fletcher, K., Corrow, S., Nair, R. Das, Barton, J. J. S., Yonas, A., and Duchaine, B. C. (2014). "A room full of strangers every day": The psychosocial impact of developmental prosopagnosia on children and their families. *J. Psychosom. Res.* doi:10.1016/j.jpsychores.2014.06.001.

Davidshofer, K. R., and Murphy, C. O. (2005). *Psychological testing: principles and applications.* 6th ed. Upper Saddle River, NJ: Pearson/Prentice Hal.

DeGutis, J. M., Chatterjee, G., Mercado, R. J., and Nakayama, K. (2012). Face gender recognition in developmental prosopagnosia: Evidence for holistic processing and use of configural information. *Vis. cogn.* 20, 1242–1253. doi:10.1080/13506285.2012.744788.

DeGutis, J. M., Cohan, S., and Nakayama, K. (2014). Holistic face training enhances face processing in developmental prosopagnosia. *Brain* 137, 1781–98. doi:10.1093/brain/awu062.

Degutis, J. M., Wilmer, J. B., Mercado, R. J., and Cohan, S. (2013). Using regression to measure holistic face processing reveals a strong link with face recognition ability. *Cognition* 126, 87–100. doi:10.1016/j.cognition.2012.09.004.

Dennett, H. W., McKone, E., Tavashmi, R., Hall, A., Pidcock, M., Edwards, M., and Duchaine, B. C. (2011). The Cambridge Car Memory Test: A task matched in format to the Cambridge Face Memory Test, with norms, reliability, sex differences, dissociations from face memory, and expertise effects. *Behav. Res. Methods*. doi:10.3758/s13428-011-0160-2.

Dobel, C., Bölte, J., Aicher, M., and Schweinberger, S. R. (2007). Prosopagnosia without apparent cause: Overview and diagnosis of six cases. *Cortex* 2, 718–733.

Dobs, K., Bülthoff, I., and Schultz, J. (2015). Identity information in facial motion varies with the type of facial movement. *Manuscr. Prep.*

Duchaine, B. C., Germine, L. T., and Nakayama, K. (2007a). Family resemblance: ten family members with prosopagnosia and within-class object agnosia. *Cogn. Neuropsychol.* 24, 419–30. doi:10.1080/02643290701380491.

Duchaine, B. C., and Nakayama, K. (2006a). Developmental prosopagnosia: a window to content-specific face processing. *Curr. Opin. Neurobiol.* 16, 166–73. doi:10.1016/j.conb.2006.03.003.

Duchaine, B. C., and Nakayama, K. (2006b). The Cambridge Face Memory Test: results for neurologically intact individuals and an investigation of its validity using inverted face stimuli and prosopagnosic participants. *Neuropsychologia* 44, 576–85. doi:10.1016/j.neuropsychologia.2005.07.001.

Duchaine, B. C., Parker, H., and Nakayama, K. (2003). Normal recognition of emotion in a prosopagnosic. *Perception* 32, 827–838. doi:10.1068/p5067.

Duchaine, B. C., Yovel, G., Butterworth, E., and Nakayama, K. (2006). Prosopagnosia as an impairment to face-specific mechanisms: Elimination of the alternative hypotheses in a developmental case. *Cogn. Neuropsychol.* 23, 714–747. doi:10.1080/02643290500441296.

Duchaine, B. C., Yovel, G., and Nakayama, K. (2007b). No global processing deficit in the Navon task in 14 developmental prosopagnosics. *Soc. Cogn. Affect. Neurosci.* 2, 104–13. doi:10.1093/scan/nsm003.

Eddy, J. K., and Glass, A. L. (1981). Reading and listening to high and low imagery sentences. *J. Verbal Learning Verbal Behav.* 20, 333–345. doi:10.1016/S0022-5371(81)90483-7.

Esins, J., Bülthoff, I., Kennerknecht, I., and Schultz, J. (2012). Can a test battery reveal subgroups in congenital prosopagnosia? *Percept. 41 ECVP Abstr. Suppl.* 41, 113–113.

Esins, J., Bülthoff, I., and Schultz, J. (2014a). Motion does not improve face recognition accuracy in congenital prosopagnosia. *J. Vis. 2014 VSS Abstr. Suppl.* 14, 1436–1436. doi:10.1167/14.10.1436.

Esins, J., Bülthoff, I., and Schultz, J. (2011). The role of featural and configural information for perceived similarity between faces. *J. Vis. 2011 VSS Abstr. Suppl.* 11, 673–673. doi:10.1167/11.11.673.

Esins, J., Schultz, J., Bülthoff, I., and Kennerknecht, I. (2013). Galactose uncovers face recognition and mental images in congenital prosopagnosia: The first case report. *Nutr. Neurosci.* 0, 1–2. doi:10.1179/1476830513Y.0000000091.

Esins, J., Schultz, J., Wallraven, C., and Bülthoff, I. (2014b). Do congenital prosopagnosia and the other-race effect affect the same face recognition mechanisms? *Front. Hum. Neurosci.* 8, 1–14. doi:10.3389/fnhum.2014.00759.

Farah, M. J., Wilson, K. D., Drain, M., and Tanaka, J. W. (1998). What is "special" about face perception? *Psychol. Rev.* 105, 482–98. Available at: http://www.ncbi.nlm.nih.gov/pubmed/9697428.

Farah, M. J., Wilson, K. D., Maxwell Drain, H., and Tanaka, J. R. (1995). The inverted face inversion effect in prosopagnosia: Evidence for mandatory, face-specific perceptual mechanisms. *Vision Res.* 35, 2089–2093. doi:10.1016/0042-6989(94)00273-O.

Fisher, R. A. (1921). On the "Probable Error" of a Coefficient of Correlation Deduced from a Small Sample. *Metron* 1, 3–32.

Freire, A., Lee, K., and Symons, L. a (2000). The face-inversion effect as a deficit in the encoding of configural information: Direct evidence. *Perception* 29, 159–170. doi:10.1068/p3012.

Garrido, L., Duchaine, B. C., and Nakayama, K. (2008). Face detection in normal and prosopagnosic individuals. *J. Neuropsychol.* 2, 119–140. doi:10.1348/174866407X246843.

Garrido, L., Furl, N., Draganski, B., Weiskopf, N., Stevens, J., Tan, G. C.-Y., Driver, J., Dolan, R. J., and Duchaine, B. C. (2009). Voxel-based morphometry reveals reduced grey matter volume in the temporal cortex of developmental prosopagnosics. *Brain* 132, 3443–55. doi:10.1093/brain/awp271.

De Gelder, B., Bachoud-Lévi, A.-C., and Degos, J.-D. (1998). Inversion superiority in visual agnosia may be common to a variety of orientation polarised objects besides faces. *Vision Res.* 38, 2855–2861. doi:10.1016/S0042-6989(97)00458-6.

Gomez, J. L., Pestilli, F., Witthoft, N., Golarai, G., Liberman, A., Poltoratski, S., Yoon, J., and Grill-Spector, K. (2015). Functionally Defined White Matter Reveals Segregated Pathways in Human Ventral Temporal Cortex Associated with Category-Specific Processing. *Neuron* 85, 216–227. doi:10.1016/j.neuron.2014.12.027.

Le Grand, R., Cooper, P. A., Mondloch, C. J., Lewis, T. L., Sagiv, N., De Gelder, B., and Maurer, D. (2006). What aspects of face processing are impaired in developmental prosopagnosia? *Brain Cogn.* 61, 139–58. doi:10.1016/j.bandc.2005.11.005.

Gruber, T., Dobel, C., Jungho, M., and Junghöfer, M. (2011). The Role of Gamma-Band Activity in the Representation of Faces: Reduced Activity in the Fusiform Face Area in Congenital Prosopagnosia. *PLoS One* 6, e19550. doi:10.1371/journal.pone.0019550.

Grüter, M. (2004). Genetik der kongenitalen Prosopagnosie [Genetics of congenital prosopagnosia]. *Unpubl. Dr. Thesis, Medizinische Fak. der Westfälischen Wilhelms-Universität Münster, Münster.*

Grüter, M., Grüter, T., Bell, V., Horst, J., Laskowski, W., Sperling, K., Halligan, P. W., Ellis, H. D., and Kennerknecht, I. (2007). Hereditary Prosopagnosia: the first case series. *Cortex* 43, 734–749. Available at: http://thomasgrueter.de/Grueter_et_al_2007cortex.pdf [Accessed January 17, 2012].

Grüter, T., Grüter, M., and Carbon, C.-C. (2011). Congenital prosopagnosia. Diagnosis and mental imagery: commentary on "Tree JJ, and Wilkie J. Face and object imagery in congenital prosopagnosia: a case series.". *Cortex.* 47, 511–3. doi:10.1016/j.cortex.2010.08.005.

Grüter, T., Grüter, M., and Carbon, C.-C. (2008). Neural and genetic foundations of face recognition and prosopagnosia. *J. Neuropsychol.* 2, 79–97. doi:10.1348/174866407X231001.

De Haan, E. H. F., and Campbell, R. (1991). A fifteen year follow-up of a case of developmental prosopagnosia. *Cortex* 27, 489–509. doi:10.1016/s0010-9452(13)80001-9.

Haxby, J. V., Hoffman, E. A., and Gobbini, M. I. (2000). The distributed human neural system for face perception. *Trends Cogn. Sci.* 4, 223–233. Available at: http://www.ncbi.nlm.nih.gov/pubmed/10827445.

Hayward, W. G., Rhodes, G., and Schwaninger, A. (2008). An own-race advantage for components as well as configurations in face recognition. *Cognition* 106, 1017–27. doi:10.1016/j.cognition.2007.04.002.

Herzmann, G., Danthiir, V., Schacht, A., Sommer, W., and Wilhelm, O. (2008). Toward a comprehensive test battery for face cognition: Assessment of the tasks. *Behav. Res. Methods* 40, 840–857. doi:10.3758/BRM.40.3.840.

Hoffman, E. A., and Haxby, J. V. (2000). Distinct representations of eye gaze and identity in the distributed human neural system for face perception. *Nat. Neurosci.* 3, 80–4. doi:10.1038/71152.

IBM Corp. Released 2011. IBM SPSS Statistics for Windows, Version 20.0. Armonk, NY: IBM Corp.

Ishai, A., Schmidt, C. F., and Boesiger, P. (2005). Face perception is mediated by a distributed cortical network. *Brain Res. Bull.* 67, 87–93. doi:10.1016/j.brainresbull.2005.05.027.

Jones, R. D., and Tranel, D. (2001). Severe Developmental Prosopagnosia in a Child With Superior Intellect. *J. Clin. Exp. Neuropsychol.* 23, 265–273(9). doi:http://dx.doi.org/10.1076/jcen.23.3.265.1183.

Kanwisher, N., and Yovel, G. (2006). The fusiform face area: a cortical region specialized for the perception of faces. *Philos. Trans. R. Soc. Lond. B. Biol. Sci.* 361, 2109–28. doi:10.1098/rstb.2006.1934.

Kaulard, K., Cunningham, D. W., Bülthoff, H. H., and Wallraven, C. (2012). The MPI facial expression database--a validated database of emotional and conversational facial expressions. *PLoS One* 7, e32321. doi:10.1371/journal.pone.0032321.

Kennerknecht, I. (2011, 9 February). Prosopagnosie oder das Problem, Gesichter wieder zu erkennen. Available at: http://www.prosopagnosia.de/ [Accessed January 8, 2015].

Kennerknecht, I., Grüter, T., Welling, B., and Wentzek, S. (2006). First Report of Prevalence of Non-Syndromic Hereditary Prosopagnosia (HPA). *Am. J. Med. Genet.*, 1617 – 1622. doi:10.1002/ajmg.a.

Kennerknecht, I., Ho, N. Y., & Wong, V. C. N. (2008a). Prevalence of hereditary prosopagnosia (HPA) in Hong Kong Chinese population. *American Journal of Medical Genetics. Part A*, *146A*(22), 2863–70. doi:10.1002/ajmg.a.32552

Kennerknecht, I., Kischka, C., Stemper, C., Elze, T., and Stollhoff, R. (2011). "Heritability of face recognition," in *Face Analysis, Modeling and Recognition Systems*, ed. T. Barbu (InTech), 163–188. Available at: http://gendocs.ru/docs/18/17804/conv_1/file1.pdf#page=175 [Accessed July 14, 2014].

Kennerknecht, I., Plümpe, N., Edwards, S., and Raman, R. (2007). Hereditary prosopagnosia (HPA): the first report outside the Caucasian population. *J. Hum. Genet.* 52, 230–6. doi:10.1007/s10038-006-0101-6.

Kennerknecht, I., Welling, B., and Pluempe, N. (2008b). Congenital prosopagnosia – a common hereditary cognitive dysfunction in humans. *Front. Biosci.* 13, 3150–3158.

Kimchi, R., Behrmann, M., Avidan, G., and Amishav, R. (2012). Perceptual separability of featural and configural information in congenital prosopagnosia. *Cogn. Neuropsychol.* 29, 447–63. doi:10.1080/02643294.2012.752723.

Kleiner, M., Brainard, D., and Pelli, D. (2007). What's new in Psychtoolbox-3? in *Perception 36 ECVP Abstract Supplement.*

Konar, Y., Bennett, P. J., and Sekuler, A. B. (2010). Holistic processing is not correlated with face-identification accuracy. *Psychol. Sci.* 21, 38–43. doi:10.1177/0956797609356508.

Kress, T., and Daum, I. (2003). Developmental prosopagnosia: a review. *Behav. Neurol.* 14, 109–21. doi:10.1155/2003/520476.

Lance, C. E., Butts, M. M., and Michels, L. C. (2006). The Sources of Four Commonly Reported Cutoff Criteria. *Organ. Res. Methods* 9, 202–220.

Lange, J., de Lussanet, M., Kuhlmann, S., Zimmermann, A., Lappe, M., Zwitserlood, P., and Dobel, C. (2009). Impairments of biological motion perception in congenital prosopagnosia. *PLoS One* 4, e7414. doi:10.1371/journal.pone.0007414.

Lee, H.-J., Macbeth, A. H., Pagani, J., and Young, W. S. 3rd (2009). Oxytocin: the Great Facilitator of Life. *Prog. Neurobiol.* 88, 127–151. doi:10.1016/j.pneurobio.2009.04.001.Oxytocin.

Lobmaier, J. S., Bölte, J., Mast, F. W., and Dobel, C. (2010). Configural and featural processing in humans with congenital prosopagnosia. *Adv. Cogn. Psychol.* 6, 23–34. doi:10.2478/v10053-008-0074-4.

Longmore, C. A., and Tree, J. J. (2013). Motion as a cue to face recognition: Evidence from congenital prosopagnosia. *Neuropsychologia* 51, 1–12. doi:10.1016/j.neuropsychologia.2013.01.022.

Malpass, R. S., and Kravitz, J. (1969). Recognition for faces of own and other race. *J. Pers. Soc. Psychol.* 13, 330–4. Available at: http://www.ncbi.nlm.nih.gov/pubmed/5359231.

Markovska-Simoska, S. M., and Pop-Jordanova, N. (2010). Face and emotion recognition by ADHD and normal adults. *Acta Neuropsychol.* 8, 99–122.

Marks, D. F. (1973). Visual imagery differences in the recall of pictures. *Br. J. Psychol.* 64, 17–24. Available at: http://www.ncbi.nlm.nih.gov/pubmed/4742442.

Maurer, D., Le Grand, R., and Mondloch, C. J. (2002). The many faces of configural processing. *Trends Cogn. Sci.* 6, 255–260. doi:10.1016/S1364-6613(02)01903-4.

Maurer, D., O'Craven, K. M., Le Grand, R., Mondloch, C. J., Springer, M. V, Lewis, T. L., and Grady, C. L. (2007). Neural correlates of processing facial identity based on features versus their spacing. *Neuropsychologia* 45, 1438–51. doi:10.1016/j.neuropsychologia.2006.11.016.

Mayer, E., and Rossion, B. (2009). "Prosopagnosia," in *The Behavioral and Cognitive Neurology of Stroke*, eds. O. Godefroy and J. Bogousslavsky (Cambridge: Cambridge University Press), 316–335. Available at: https://www.yumpu.com/en/document/view/12830258/prosopagnosia.

McKone, E., Davies, A. A., Darke, H., Crookes, K., Wickramariyaratne, T., Zappia, S., Fiorentini, C., Favelle, S., Broughton, M., and Fernando, D. (2013). Importance of the inverted control in measuring holistic face processing with the composite effect and part-whole effect. *Front. Psychol.* 4, 33. doi:10.3389/fpsyg.2013.00033.

McKone, E., Hall, A., Pidcock, M., Palermo, R., Wilkinson, R. B., Rivolta, D., Yovel, G., Davis, J. M., and O'Connor, K. B. (2011). Face ethnicity and measurement reliability affect face recognition performance in developmental prosopagnosia: evidence from the Cambridge Face Memory Test-Australian. *Cogn. Neuropsychol.* 28, 109–46. doi:10.1080/02643294.2011.616880.

McKone, E., Stokes, S., Liu, J., Cohan, S., Fiorentini, C., Pidcock, M., Yovel, G., Broughton, M., and Pelleg, M. (2012). A robust method of measuring other-race and other-ethnicity effects: the Cambridge Face Memory Test format. *PLoS One* 7, 1–6. doi:10.1371/journal.pone.0047956.

Meissner, C. A., and Brigham, J. C. (2001). Thirty years of investigating the own-race bias in memory for faces: A meta-analytic review. *Psychol. Public Policy, Law* 7, 3–35. doi:10.1037//1076-8971.7.1.3.

Michel, C., Caldara, R., and Rossion, B. (2006a). Same-race faces are perceived more holistically than other-race faces. *Vis. cogn.* 14, 55–73. doi:10.1080/13506280500158761.

Michel, C., Rossion, B., Han, J., Chung, C.-S., and Caldara, R. (2006b). Holistic processing is finely tuned for faces of one's own race. *Psychol. Sci.* 17, 608–15. doi:10.1111/j.1467-9280.2006.01752.x.

Mosetter, K. (2008). "Chronischer Streß auf der Ebene der Molekularbiologie und Neurobiochemie," in *Psychodynamische Psycho-und Traumatherapie* (VS Verlag für Sozialwissenschaften), 77–98.

Nieminen-von Wendt, T., Paavonen, J. E., Ylisaukko-Oja, T., Sarenius, S., Källman, T., Järvelä, I., and von Wendt, L. (2005). Subjective face recognition difficulties, aberrant sensibility, sleeping disturbances and aberrant eating habits in families with Asperger syndrome. *BMC Psychiatry* 5, 20. doi:10.1186/1471-244X-5-20.

Nunnally, J. C. (1970). *Introduction to psycological measurement.* New York: McGraw-Hill.

Nunnally, J. C. (1978). *Psychometric theory.* 2nd ed. New York: McGraw-Hill.

O'Toole, A. J., Harms, J., Snow, S. L., Hurst, D. R., Pappas, M. R., Ayyad, J. H., and Abdi, H. (2005). A video database of moving faces and people. *IEEE Trans. Pattern Anal. Mach. Intell.* 27, 812–6. doi:10.1109/TPAMI.2005.90.

Palermo, R., Willis, M. L., Rivolta, D., McKone, E., Wilson, C. E., and Calder, A. J. (2011). Impaired holistic coding of facial expression and facial identity in congenital prosopagnosia. *Neuropsychologia* 49, 1226–35. doi:10.1016/j.neuropsychologia.2011.02.021.

Pitcher, D., Walsh, V., and Duchaine, B. C. (2011). The role of the occipital face area in the cortical face perception network. *Exp. brain Res.* 209, 481–93. doi:10.1007/s00221-011-2579-1.

Pyles, J. a, Verstynen, T. D., Schneider, W., and Tarr, M. J. (2013). Explicating the face perception network with white matter connectivity. *PLoS One* 8, e61611. doi:10.1371/journal.pone.0061611.

Rhodes, G., Brake, S., Taylor, K., and Tan, S. (1989). Expertise and configural coding in face recognition. *Br. J. Psychol.* 80, 313–331. doi:10.1111/j.2044-8295.1989.tb02323.x.

Richler, J. J., Cheung, O. S., and Gauthier, I. (2011). Holistic processing predicts face recognition. *Psychol. Sci.* 22, 464–471. doi:10.1177/0956797611401753.Holistic.

Rivolta, D., Palermo, R., Schmalzl, L., and Coltheart, M. (2011). Covert face recognition in congenital prosopagnosia: A group study. *Cortex* 48, 1–9. doi:10.1016/j.cortex.2011.01.005.

Rivolta, D., Palermo, R., Schmalzl, L., and Williams, M. a (2012). Investigating the features of the m170 in congenital prosopagnosia. *Front. Hum. Neurosci.* 6, 45. doi:10.3389/fnhum.2012.00045.

Rodrigues, A., Bolognani, S. A. P. S., Brucki, S. S. M. D., and Bueno, O. F. A. O. (2008). Developmental prosopagnosia and adaptative compensatory strategies. *Dement. Neuropsychol.* 2, 353–355. Available at: http://www.demneuropsy.com.br/imageBank/PDF/dnv02n04a20.pdf [Accessed September 29, 2014].

Russell, R., Duchaine, B. C., and Nakayama, K. (2009). Super-recognizers: people with extraordinary face recognition ability. *Psychon. Bull. Rev.* 16, 252–7. doi:10.3758/PBR.16.2.252.

Salkovic-Petrisic, M., Osmanovic-Barilar, J., Knezovic, A., Hoyer, S., Mosetter, K., and Reutter, W. (2014). Long-term oral galactose treatment prevents cognitive deficits in male Wistar rats treated intracerebroventricularly with streptozotocin. *Neuropharmacology* 77, 68–80. doi:10.1016/j.neuropharm.2013.09.002.

Savaskan, E., Ehrhardt, R., Schulz, A., Walter, M., and Schächinger, H. (2008). Post-learning intranasal oxytocin modulates human memory for facial identity. *Psychoneuroendocrinology* 33, 368–74. doi:10.1016/j.psyneuen.2007.12.004.

Schmalzl, L., Palermo, R., and Coltheart, M. (2008a). Cognitive heterogeneity in genetically based prosopagnosia: A family study. *J. Neuropsychol.* 2, 99–117. doi:10.1348/174866407X256554.

Schmalzl, L., Palermo, R., Green, M., Brunsdon, R., and Coltheart, M. (2008b). Training of familiar face recognition and visual scan paths for faces in a child with congenital prosopagnosia. *Cogn. Neuropsychol.* 25, 704–29. doi:10.1080/02643290802299350.

Schwarzer, G., Huber, S., and Dümmler, T. (2005). Gaze behavior in analytical and holistic face processing. *Mem. Cognit.* 33, 344–354. doi:10.3758/BF03195322.

Schweich, M., and Bruyer, R. (1993). Heterogeneity in the cognitive manifestations of prosopagnosia: The study of a group of single cases. *Cogn. Neuropsychol.* 10, 529–547. doi:10.1080/02643299308253472.

Shah, P., Gaule, A., Gaigg, S. B., Bird, G., and Cook, R. (2015). Probing short-term face memory in developmental prosopagnosia. *Cortex* 64, 115–122. doi:10.1016/j.cortex.2014.10.006.

Spearman, C. (1910). Correlation calculated from faulty data. *Br. J. Psychol. 1904-1920* 3, 270–295.

Stollhoff, R. (2010). Modeling Prosopagnosia: Computational Theory and Experimental Investigations of a Deficit in Face Recognition (Doctoral dissertation, Phd thesis). Available at: http://www.fmi.unileipzig.de/promotion/abstract.stollhoff.pdf.

Stollhoff, R., Jost, J., Elze, T., and Kennerknecht, I. (2011). Deficits in long-term recognition memory reveal dissociated subtypes in congenital prosopagnosia. *PLoS One* 6, e15702. doi:10.1371.

Susilo, T., McKone, E., Dennett, H. W., Darke, H., Palermo, R., Hall, A., Pidcock, M., Dawel, A., Jeffery, L., Wilson, C. E., et al. (2010). Face recognition impairments despite normal holistic processing and face space coding: Evidence from a case of developmental prosopagnosia. *Cogn. Neuropsychol.* 27, 636–664. doi:10.1080/02643294.2011.613372.

Tanaka, J. W., and Farah, M. J. (1993). Parts and wholes in face recognition. *Q. J. Exp. Psychol. A.* 46, 225–45. Available at: http://www.ncbi.nlm.nih.gov/pubmed/8316637.

The MathWorks Inc. MATLAB and Statistics Toolbox Release 2011b. Natick, Massachusetts, United States.

Thomas, C., Avidan, G., Humphreys, K., Jung, K.-J., Gao, F., and Behrmann, M. (2009). Reduced structural connectivity in ventral visual cortex in congenital prosopagnosia. *Nat. Neurosci.* 12, 29–31. doi:10.1038/nn.2224.

Tree, J. J., and Wilkie, J. (2010). Face and object imagery in congenital prosopagnosia: a case series. *Cortex* 46, 1189–98. doi:10.1016/j.cortex.2010.03.005.

Wilmer, J. B., Germine, L., Chabris, C. F., Chatterjee, G., Williams, M., Loken, E., Nakayama, K., and Duchaine, B. C. (2010). Human face recognition ability is specific and highly heritable. *Proc. Natl. Acad. Sci. U. S. A.* 107, 5238–41. doi:10.1073/pnas.0913053107.

Xiao, N. G., Quinn, P. C., Ge, L., and Lee, K. (2013). Elastic facial movement influences part-based but not holistic processing. *J. Exp. Psychol. Hum. Percept. Perform.* 39, 1457–67. doi:10.1037/a0031631.

Young, A. W., Hellawell, D., and Hay, D. C. (1987). Configurational information in face perception. *Perception* 16, 747–759. Available at: http://www.perceptionweb.com/perception/fulltext/p16/p160747.pdf [Accessed February 20, 2014].

Yovel, G., and Duchaine, B. C. (2006). Specialized face perception mechanisms extract both part and spacing information: evidence from developmental prosopagnosia. *J. Cogn. Neurosci.* 18, 580–93. doi:10.1162/jocn.2006.18.4.580.

Zhao, M., Hayward, W. G., and Bülthoff, I. (2014). Holistic processing, contact, and the other-race effect in face recognition. *Vision Res.* 105, 61–69. doi:10.1016/j.visres.2014.09.006.

Declaration of Contribution

This thesis comprises three manuscripts that are either published or prepared for publication. Details about these manuscripts are presented in the following.

The candidate developed the ideas for the studies, their experimental design and implementation in collaboration with the supervisors, and recruited participants, collected and analyzed the data. The co-authors supervised the work of the candidate and assisted in the revision of the manuscripts.

1. Esins, J., Schultz, J., Stemper C., Kennerknecht, I. & Bülthoff, I. (2015). Face perception and test reliabilities in congenital prosopagnosia in seven tests. (prepared for submission): J.E. created the stimuli, programmed the experiment, collected and analyzed the data. All authors conceived the set-up of the assessment and wrote the manuscript.

2. Esins, J., Schultz, J., Wallraven, C., & Bülthoff, I. (2014). Do congenital prosopagnosia and the other-race effect affect the same face recognition mechanisms? *Frontiers in Human Neuroscience*, *8*(September), 1–14. doi:10.3389/fnhum.2014.00759; and Esins J, Schultz J, Stemper C, Kennerknecht I, Wallraven C and Bülthoff I (2015). Corrigendum: Do congenital prosopagnosia and the other-race effect affect the same face recognition mechanisms?. Front. Hum. Neurosci. 9:294. doi: 10.3389/fnhum.2015.00294: J.E. created the stimuli, programmed the experiment, recruited the German control participants, collected data of Caucasian participants and analyzed the data. C.S and I.K. recruited the prosopagnisc participants. Recruitment and data collection of the Korean participants was conducted by Bora Kim, Department of Brain and Cognitive Engineering, Korea University, Seoul, South Korea. All authors conceived the set-up of the assessment and wrote the manuscript.

3. Esins, J., Schultz, J., Bülthoff, I., & Kennerknecht, I. (2013). Galactose uncovers face recognition and mental images in congenital prosopagnosia: The first case report. *Nutritional Neuroscience*, *0*(0), 1–2. doi:10.1179/1476830513Y.0000000091:

I.K. was first contacted and informed about the effects of galactose by the participant. J.E. conducted the interviews with the participant and the assessment of the influence of galactose on further prosopagnosics. All authors conceived the set-up of the assessment and wrote the manuscript .

Parts of this work were also presented at the following conferences:

1. Esins, J., Bülthoff, I., & Schultz, J. (2011, May). The role of featural and configural information for perceived similarity between faces. Poster presented at the 11th Annual Meeting of the Vision Sciences Society: VSS 2011, Naples, FL, USA. Journal of Vision, 11(11), 673-673, doi:10.1167/11.11.673.

2. Esins, J., Schultz, J., Kim, B., Wallraven, C., & Bülthoff, I. (2012). Comparing the other-race-effect and congenital prosopagnosia using a three-experiment test battery. Poster presented at the Asia-Pacific Conference on Vision 2012: APCV 2012, Incheon, South Korea. i-Perception, 3(9), 688-688.

3. Kim, B., Esins, J., Schultz, J., Bülthoff, I., & Wallraven, C. (2012). Mapping the other-race-effect in face recognition using a three-experiment test battery. Poster presented by B.K. at the Asia-Pacific Conference on Vision 2012: APCV 2012, Incheon, South Korea. i-Perception, 3(9), 711-711.

4. Esins, J., Bülthoff, I., Kennerknecht, I., & Schultz, J. (2012). Can a test battery reveal subgroups in congenital prosopagnosia? Poster presented at the 35th European Conference on Visual Perception: ECVP 2012, Alghero, Italy. Perception, Vol. 41, 113-113.

5. Esins, J., Bülthoff, I., & Schultz, J. (2014). Motion does not improve face recognition accuracy in congenital prosopagnosia. Poster presented at the 14th Annual Meeting of the Vision Sciences Society: VSS 2014, Naples, FL, USA. Journal of Vision, 14(10), 1436-1436, doi:10.1167/14.10.1436.

II. Face perception and test reliabilities in congenital prosopagnosia in seven tests

1. Abstract

Congenital prosopagnosia, the innate impairment in recognizing faces, is a very heterogeneous disorder with different forms of manifestation. To investigate the nature of prosopagnosia in more detail, we tested 16 prosopagnosics and 21 matched controls with an extended test battery, addressing various aspects of face recognition. Our results show that on a group-wise level prosopagnosics showed significant impairments in several face recognition tasks: they showed impaired holistic processing (amongst others tested with the Cambridge Famous Face Test (CFMT)), as well as impaired processing of configural information of faces. While controls showed an improvement of face recognition accuracy for moving compared to static faces, prosopagnosics did not show this effect. Furthermore, prosopagnosics showed a significantly impaired gender recognition. There was no difference between groups in the automatic extraction of face identity information, or in object recognition as tested with the Cambridge Car Memory Test.

In addition, a methodological analysis of the tests revealed reduced internal consistency reliability for holistic face processing tests in prosopagnosics. In particular, prosopagnosics showed a significantly reduced reliability coefficient (Cronbach's alpha) in the CFMT compared to the controls. We suggest that feature-based response strategies, employed by the prosopagnosics, might cause a heterogeneous response pattern, which is revealed by the reduced test reliability. This finding raises the question whether classical face tests measure the same perceptual processes in controls and prosopagnosics.

2.   Introduction

Congenital prosopagnosia refers to the lifelong, innate impairment in identifying someone by his or her face (Bodamer, 1947). It is estimated to affect about 2.5% of the population (Kennerknecht et al., 2006), and is characterized as a neurodevelopmental disorder of face recognition without any deficits in low-level vision or intelligence (Behrmann and Avidan, 2005). Not only face identification was found to be impaired in congenital prosopagnosia, but also other aspects of face processing. However, psychophysical studies differ in their findings. For example, some studies found impaired gender recognition in congenital prosopagnosics (Duchaine and Nakayama, 2006a; Ariel and Sadeh, 1996), while others reported gender recognition to be normal (Chatterjee and Nakayama, 2012). The same is true for sensitivity to featural and configural facial information, for which no uniform pattern of impairments was found (Yovel and Duchaine, 2006; Le Grand et al., 2006). There is also disagreement over holistic processing. While Avidan and colleagues found evidence for weaker holistic processing when testing 14 prosopagnosics (Avidan et al., 2011), Le Grand and colleagues reported that only one of their eight prosopagnosic participants showed reduced holistic processing (Le Grand et al., 2006). In short, the picture of a very heterogeneous disorder emerges from these equivocal results (Schweich and Bruyer, 1993; Le Grand et al., 2006), even when accounting for the differences in experiment and stimulus design.

For these reasons, we tested face perception in congenital prosopagnosia with a rather large sample of prosopagnosics (16) in more detail. We included two widely used tests for reference, the Cambridge Face Memory test (CFMT, Duchaine and Nakayama, 2006b) and the Cambridge Car Memory Test (CCMT, Dennett et al., 2011). We also used test paradigms for which some controversial results exist in literature, and we developed new tests to investigate perceptual processes in congenital prosopagnosia in more depth.

In the present study, we report and compare the performance of a group of 16 congenital prosopagnosics to the performance of 21 matched controls in seven tests. We tested holistic face processing, configural and featural face processing, processing of faces in motion, unconscious processing of faces, face gender recognition, and object recognition. For each test separately, we will present motivation, methodological details, results and discussion. In addition, we calculated test reliabilities for each participant group, which will be discussed in a separate section, followed by the general discussion.

3. General methods

3.1. Procedure

The experiments were conducted in two sessions lying about two years apart: on average 24.6 months (SD = 2.3) for prosopagnosics and 20.3 months (SD = 1.6) for controls. During the first session, participants performed the Cambridge Face Memory Test (CFMT), a surprise recognition test and a similarity rating test. The second session included the Cambridge Car Memory Test (CCMT), a facial motion advantage test, the composite face test, and a gender recognition test. In both sessions, participants could take self-paced breaks between the experiments.

All participants were tested individually. The experiments were run on a desktop PC with 24" screen. The CFMT and CCMT are Java-script based; the other experiments were run with Matlab2011b (The MathWorks Inc.) and Psychtoolbox (Brainard, 1997; Kleiner et al., 2007). Participants were seated at a viewing distance of approximately 60 cm from the screen. The procedure was approved by the local ethics committee.

3.2. Participants

We tested 16 congenital prosopagnosic participants (from now on referred to as "prosopagnosics") and 21 control participants ("controls"). All participants provided informed consent. All participants had normal or corrected-to-normal visual acuity.

Prosopagnosics

The prosopagnosics were diagnosed by the Institut für Humangenetik, Westfälische-Wilhelms-Universität, Münster, Germany, based on a screening questionnaire and a diagnostic semi-structured interview (Stollhoff et al., 2011). All prosopagnosics were tested at the Max Planck Institute for Biological Cybernetics in Tübingen, Germany, and compensated with 8 Euro per hour plus travel expenses.

35

Controls

All controls were tested at the Max Planck Institute for Biological Cybernetics in Tübingen, Germany, and compensated with 8 Euro per hour. Controls were chosen to fit as closely as possible the prosopagnosic participants in terms of age, sex and schooling level. Due to time constraints, the controls did not participate in the diagnostic interview but reported to have no problems in recognizing faces of their friends and family members.

Table 1: Participants' demographics

	Prosopagnosics		Controls	
	Sex	Age	Sex	Age
1	f	22	f	21
2	f	24	f	24
3	f	27	f	24
4	f	28	f	28
5	m	33	f	29
6	m	34	f	31
7	f	36	m	33
8	m	36	m	36
9	m	37	m	37
10	f	41	f	37
11	f	46	m	38
12	m	47	m	39
13	m	52	m	39
14	f	54	f	42
15	m	57	m	44
16	m	59	f	44
17			f	47
18			m	48
19			f	49
20			f	58
21			m	60
♂	8		9	
Mean age		39.6		38.5

Age (in years) and gender ("m" = male, "f" = female) of prosopagnosic and control participants

To provide an objective measure of face processing abilities and to maintain comparability with other studies, we tested all participants with the Cambridge Face Memory Test (CFMT). The z-scores are given in Table 2 for prosopagnosics and Table 3 for controls. These tables also contain the z-scores of all other tests reported in this study. Z-scores for both groups were calculated based on the results of the control participants.

3.3. Analysis

The description of the dependent variables is given for each test individually. All analyses were conducted with Matlab2011b (The MathWorks Inc.) and IBM SPSS statistics Version 20 (IBM Corp. Released 2011.). ANOVASs and their effect sizes (η^2) and linear regressions were calculated with IBM SPSS statistics Version 20. T-tests and their effect sizes Cohen's d (d), as well as Mann-Whitney U-tests were calculated with Matlab2011b. Tests' internal consistency reliability coefficients were calculated with Matlab2011b.

Where possible, tests reliability was calculated as Cronbach's alpha with the function cronbach.m for Matlab, written by Alexandros Leontitsis, Department of Education, University of Ioannina, Greece. Furthermore, we calculated reliability with the split-half method and subsequent adjustment with the Spearman–Brown prediction formula for all tests: The trials of a test are split into halves (e.g. first half versus second half, or odd trials versus even trials). Then the mean score of each half is calculated for each participant. The correlation between participant's mean half scores gives an estimate of the test reliability (Davidshofer and Murphy, 2005). We adapted this method by bootstrapping: test trials were split randomly into halves, followed by correlation of the mean half scores. This procedure was repeated 100,000 times. The median of these bootstrapped correlations was then adjusted to the tests full length with the Spearman–Brown prediction formula (Brown, 1910; Spearman, 1910).

Statistical difference between reliability coefficients for Cronbach's alpha was calculated based on the Fisher-Bonett approach (Bonett, 2003 [1]; formula (2)). Statistical difference between adapted split-half reliability coefficients was calculated as statistical difference between correlation coefficients (Fisher, 1921) for the uncorrected reliability coefficients (i.e. before applying the Spearman–Brown prediction formula).

---

[1] Please note the typo in formula (2) for this reference. It should read as ... {(k1 - 1)(n1 - 2))}...

4. Tests

4.1. Cambridge Face Memory Test

Motivation

The Cambridge Face Memory Test (CFMT) was created and provided by Duchaine and Nakayama (2006b). It is a widely used test to characterize prosopagnosic participants (Kimchi et al., 2012; Rivolta et al., 2011) and to assess holistic recognition abilities of unfamiliar faces. The CFMT has been confirmed to have a high internal consistency reliability with a Cronbach's alpha between .8 and .9 in different studies (Herzmann et al., 2008; Wilmer et al., 2010; Bowles et al., 2009). We used this test as an objective measure of face recognition abilities of our participants, expecting reduced recognition abilities for the prosopagnosics, and to allow comparison with other studies.

Stimuli and Task

As this test has been described in detail in the original study (Duchaine and Nakayama, 2006b), only a short description is given here. Portraits of male Caucasians serve as stimuli. The participants were familiarized with six target faces, which they then had to recognize among distractor faces in a 3-alternative-forced-choice task. Difficulty increased stepwise during the test by changing viewpoints and lighting conditions and adding noise. No feedback was given. The test can be run in an upright and inverted condition. We only used the upright condition. In our setting, the stimuli faces had a visual angle of 5.7 degrees horizontally and vertically.

Results

We calculated the overall recognition performance as the percentage of correctly recognized faces per participant. Figure 1 depicts the mean scores per group. Controls correctly recognized 81.0 % (SD = 9.4) of the test faces, while prosopagnosics scored 54.8 %, (SD = 5.9). The difference between groups was significant (one-way ANOVA: $F(1, 36) = 94.7$, $p < .001$, $\eta^2 = .73$), with prosopagnosics performing worse than controls.

Discussion

Prosopagnosics showed a significantly reduced face recognition ability compared to controls. This result reflects the impaired holistic face processing and face memory of prosopagnosic participants and replicates findings of many previous studies (Duchaine et al., 2007a; Rivolta et al., 2012; Bate et al., 2013).

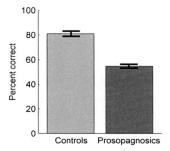

Figure 1: Mean percentage of correctly recognized faces in the CFMT for controls and prosopagnosics. Error bars: SEM.

4.2.   Cambridge Car Memory Test

Motivation

The Cambridge Car Memory Test (CCMT, Dennett et al., 2011) is a test equal in format and structure to the CFMT. We used the CCMT to test for potential general object recognition deficits. We did not expect to find recognition deficits for prosopagnosics in this control task, as only few prosopagnosics might show object recognition deficits, which are less severe than their face recognition deficits (Kress and Daum, 2003; Le Grand et al., 2006).

Dennett and colleagues found a significant correlation between the scores of their CCMT and participants' general interest in cars and knowledge of car makes and models. Therefore, we ran an additional test for car expertise after completing the CCMT, to be able to account for this possible influence and correct the CCMT scores for car expertise.

Stimuli and Task -- CCMT

As a detailed description is given in the original study (Dennett et al., 2011), we give only a short description here. The experimental design is similar to the CFMT, with pictures of whole cars serving as stimuli. The participants were familiarized with six target cars, which they then had to recognize among distractor cars in a 3-alternative-forced-choice task. Difficulty increases stepwise during the test by changing viewpoints and lighting conditions and adding noise. No feedback was given. The test can be run in an upright and inverted condition. We only used the upright condition.

Stimuli and Task -- Car expertise

Sixteen cars from the CCMT (four target and twelve distractor cars) were presented one after the other to the participants along with three answer choices of possible car makes and models (see Figure 2). Participants had to indicate the correct answer by pressing the relevant keys on the keyboard. The next image appeared as soon as a response was entered. No feedback was given and no time restrictions were applied.

The car images in both tests had a visual angle of 5.7 degrees horizontally and 11.4 degrees vertically.

Figure 2: Example trial of the car expertise test. Participants had to pick the correct answer among three written car names.

Results

The performance measure in both tasks was the percentage of correctly recognized cars per participant. Figure 3 depicts the mean scores per group and task. For the CCMT the control participants correctly recognized 77.5 % (SD = 12.9) of the cars, and prosopagnosics scored 75.1 % (SD = 12.7). For the car expertise test, controls correctly identified 64.6 % (SD = 16.1) of the car makes, and prosopagnosics scored 52.0 % (SD = 19.9).

For the CCMT there was no significant difference in scores between prosopagnosics and controls (one-way-ANOVA, $F(1, 36) = 0.31$, $p = .58$, $\eta^2 = .01$). For the car expertise test the control group exhibited significantly greater expertise in car models than the prosopagnosics ($F(1, 36) = 4.57$, $p = .04$, $\eta^2 = .12$). Therefore, we compared both groups'

CCMT scores while controlling for the car expertise. For this, we ran a linear regression with car expertise scores as predictor. The residuals of the regression did not differ significantly between groups (one-way-ANOVA, $F(1, 36) = 0.64$, $p = .43$, $\eta^2 = .02$), indicating that the CCMT scores do not differ between groups after controlling for car expertise. (Combination of both groups' regression model was possible, as groups' regression coefficients were not significantly different from each other ($t(35) = -0.33$, $p = .75$, $d = -0.11$)).

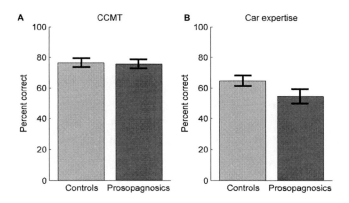

Figure 3: (A) Mean percentage of correctly recognized cars in the CCMT for controls and prosopagnosics. Error bars: SEM. (B) Mean percentage of correctly identified cars models for controls and prosopagnosics. Error bars: SEM.

Discussion

We found no difference in car recognition performance on the CCMT between controls and prosopagnosics on the group-wise level. This replicates findings of previous studies (McKone et al., 2011; Shah et al., 2015) Even though our control group contained significantly more car experts, we also could not find significant differences in the CCMT scores between controls and prosopagnosics, after correcting for car expertise. Furthermore, given the fact that no prosopagnosic scored less than 1.81 SD below the mean recognition performance of controls for the CCMT (see Table 2), there was no indication that our prosopagnosic participants had general object recognition deficits.

4.3. Surprise recognition task

Motivation

Because of their difficulty at recognizing faces, prosopagnosics rely on alternative feature-based strategies to identify people. They report using voice, hairdo, blemishes, or individual forms of face features (Dalrymple et al., 2014; Mayer and Rossion, 2009; Grüter et al., 2011), and use similar strategies in face recognition tasks in laboratory conditions (Duchaine et al., 2003). We developed a test designed to try to bypass these strategies. In the first part of our test, participants were first asked to name facial expressions performed by various actors (implicit learning phase), thus directing their focus to the facial expressions rather than to the identity of the actors. Afterwards, participants had to complete a surprise recognition task of the actors' faces. Thus, at test we expected to measure their holistic face recognition abilities without the interference of their usual, feature-based strategies. This first part was followed by a second, control part with a similar paradigm, but with the difference that participants knew that a holistic face recognition test would follow the presentation of the facial expressions (explicit learning phase). If prosopagnosics do not engage their usual feature-based strategies to remember the faces during the implicit learning phase (first part) but only do so during the explicit learning phase (second part), we expect them to show better performance at test after the explicit learning. More importantly, we expect prosopagnosics to exhibit a stronger recognition improvement between the two test parts than the control group, because then prosopagnosics can actively use their feature-based strategies to compensate their impaired holistic processing, while we expect controls to engage holistic processing in both parts.

Stimuli

The stimuli were derived from videos from our Max Planck facial expression database (Kaulard et al., 2012). The database consists of videos of male and female actors performing different emotional and conversational facial expressions (e.g. disgust, considering, being annoyed, etc.) without speaking. Frames extracted from one of the expression videos are shown in Figure 4.

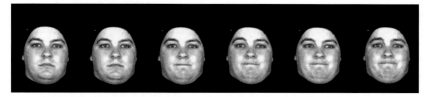

Figure 4: Some consecutive frames of a video of an actor showing the facial expression 'I don't know'

One set of 16 videos was used for the implicit learning phase and another set for the explicit learning phase. In each set, four different target actors (2 male and 2 female) were depicted, each showing four different facial expressions. Both the exhibited expressions and the actors' identities differed in both sets. The videos had a mean length of 2.7 sec (SD = 1.5). In each test phase, we used 16 static images of the target actors (see Figure 5). These images were taken from a new set of videos not presented to the participants before. As distractors we used 16 static images taken from 16 new videos with new actors (4 images each for 2 male and 2 female distractors). All videos and images were frontal views of the faces and had a visual angle of 4.8 degrees horizontally and 6.7 degrees vertically.

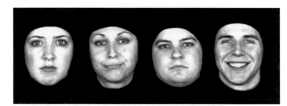

Figure 5: Example stimuli for the test phase: Static images used for testing the participants after training with dynamic videos.

Task

In the first part, during the implicit learning phase, participants saw 16 videos: 4 target actors (2 male and 2 female), each performing 4 different facial expressions that participants had to name. The order of the videos was pseudorandom such that no actor was seen twice in a row. Participants had to start each video per key press and could watch it only once. After each video, they typed in their interpretation of the facial expression (max. 80 characters). No feedback was given. After this implicit learning phase, participants performed a surprise old-new recognition task. For this, the participants saw 32 different images: four images from each of the four target actors and four images from four new distractor actors. Participants had to decide for each image whether the actor had been seen during the learning phase or not by pressing the relevant keys on the keyboard. Stimuli were presented for 2 s or until key press, whichever came first. The next image appeared as soon as an answer was entered. The order of the pictures was pseudorandom, such that no actor was seen twice in a row. No feedback was given. All participants reported that they had not anticipated the surprise recognition task after the expression naming.

The second part was conducted to control for the effect of surprise. The design was similar, with the difference that participants knew that an old-new recognition task would follow the explicit learning phase. Again, the participants watched 16 videos of four different actors. This time they did not need to name the facial expressions but could concentrate on remembering the appearance of the actors. Afterwards they once more had to recognize the actors among the distractors. Different expressions and actors were shown in the first and second part to avoid interference. The assignment of the targets and distractors to the first or second part of the experiment was randomized across participants.

Results

For each participant we calculated the d'-scores as Z(hits) − Z(false alarms). Figure 6.A depicts the mean scores per group. Controls achieved a mean d'-score of 2.09 (SD = 0.88) in the first, surprise part and 2.66 (SD = 0.91) in the second part. Prosopagnosics achieved a mean d'-score of 1.03 (SD = 0.64) in the first part and 1.48 (SD = 0.87) in the second part. A two-way repeated measures ANOVA of the factors participant group (prosopagnosics, controls) and test part (first, second) was conducted on the d'-scores. Recognition performance was significantly higher in the second part compared to the first, surprise part ($F_{(1, 35)} = 7.1$ $p = .012$, $\eta^2 = .17$) and controls performed significantly better than prosopagnosics ($F_{(1, 35)} = 29.9$, $p < .001$, $\eta^2 = .46$). The interaction between parts and participant groups was not significant ($F_{(1, 35)} = 0.11$, $p = .75$, $\eta^2 < .01$).

Prosopagnosics and controls performed significantly above chance level (prosopagnosics for both parts $t(15) > 6.4$, $p < .001$, $d > 1.61$; controls for both parts $t(20) > 10.8$, $p < .001$, $d > 2.48$). However, ceiling effects were present for the controls in the second part, as

33% of the controls scored above 95% accuracy ($\leq$ 1 error, d'-score $\geq$ 3.39), 52.4% scored above 90% accuracy ($\leq$ 3 errors, d'-score $\geq$ 2.68)), see Figure 6.B.

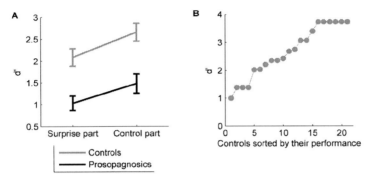

Figure 6: (A) Mean d'-scores in the surprise recognition task for controls and prosopagnosics. Error bars: SEM. (B) Ceiling effects for the control participants in the second part of the surprise recognition task.

Discussion

Overall, controls discriminated between old and new faces significantly better than the prosopagnosics in both parts. Importantly, both groups improved their performance similarly in the second part as shown by the absence of an interaction. This finding indicates that, contrary to our prediction, prosopagnosics did not exhibit a stronger recognition improvement between the two test parts compared to controls (e.g. by adapting their strategy). The ceiling performance of the controls reinforces this observation: the ceiling performance may have led to underestimate the improvement between test parts for controls, yet the improvement for prosopagnosics between test parts was still not bigger than for the controls.

Because prosopagnosics' performance was significantly above chance level in the first, surprise part, we conclude that they extracted and stored identity-relevant information even when not paying attention to that information. We suggest two equally possible explanations. First, contrary to our hypothesis, prosopagnosics had engaged their feature-based strategies not only during the explicit but also during the implicit learning phase. They were thus able to extract and store featural characteristics even without conscious effort. Second, prosopagnosics' recognition system does not differ fundamentally from that of the controls in so far as that in both groups the mechanisms of extracting identification-relevant information seem to occur automatically in explicit as well as in implicit learning conditions. These automatic mechanisms (potentially holistic processing abilities) are exhibited by prosopagnosics, yet are reduced compared to controls. In our next experiment, we investigate whether indeed holistic processing abilities are still present in reduced form in prosopagnosics.

4.4. Composite face task

Motivation

Several studies state that the key to well-functioning face recognition lies in holistic face processing. Holistic processing is defined as the integration of all facial information, e.g. shape of nose, mouth, and eyes (features) and their spatial distances (configuration). This information is combined into a whole gestalt, making it harder to process the information individually (Maurer et al., 2002). A classical test for holistic processing is the composite face task. When the top half of one face is combined with the bottom half of a different face, both halves are merged into a new, third identity. The combined face halves are processed holistically as a whole, making it difficult to retrieve the identity of the halves individually. This effect disappears when the halves are misaligned. In the composite face task participants have to indicate if one half (mostly the top half containing the eyes) is the same in two, consecutively shown composite faces. As the lower half interferes with the perception of the upper half, controls make more mistakes when the halves are aligned than when they are misaligned. This effect can also be modulated by the choice of the bottom halves: controls make more mistakes when the bottom halves are incongruent to the top halves (i.e. top halves are identical and bottom halves differ and vice versa) than for the congruent case (i.e. either top halves are identical and bottom halves are identical, or top halves differ and bottom halves differ). Our expectations were that in this task evidence of holistic processing would be generally weaker for prosopagnosics than for controls.

Stimuli

The stimuli were created from 12 images of female faces taken from the Max Planck 3D face database. All images were grey-scale and luminance equalized, so that the upper and lower half of different faces could be combined without obvious color or luminance differences. To create the composite faces, the faces were cut into top and bottom parts along the center of the image. Bottom and upper halves were rearranged according to the design of the experiment described below. The composite faces were surrounded with an oval, black mask to cover differences in the outer face shape. Moreover, a horizontal, 2 pixels thick, black line covered the border between the two halves (see Figure 7). The faces were presented with a visual angle of 2.9 degrees horizontally and 3.8 degrees vertically.

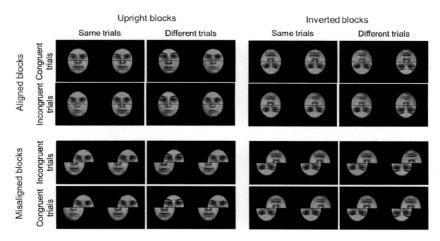

Figure 7: Example stimuli of the composite face task.

In each trial two composite faces were presented sequentially for 0.3 s each with an inter-stimulus interval of 0.4 s. The inter-trial interval was 2 s, resulting in an overall trial length of 3 s. When no face was presented, a fixation cross was shown at the center of the image. Participants were instructed to keep their gaze at the position of the fixation cross all the time, even when a face was presented and the cross was not visible.

For the 'same' condition, the top half (comprising the eyes) of the first composite face was the same as the top half of the second face within the same trial. In the 'different' condition, the two top halves differed. In the congruent condition, the bottom halves were same if the top halves were same or they were different if the top halves were different. In the incongruent condition, the bottom halves were different if the top halves were the same and vice versa. In the aligned condition, top and bottom halves were placed exactly on top of each other. For the misaligned condition, the top half was displaced 18 pixels to the right, while the bottom part was displaced 18 pixels to the left. All face images were presented upright for the upright condition or rotated by 180° for the inverted condition.

The combination of upright and inverted, and aligned and misaligned conditions was tested in four separate blocks. The block order was balanced across participants. Each of the four blocks contained 120 trials: 30 trials of each combination of same and different trials, and congruent and incongruent trials. The order of trials was randomized.

We used the "complete design" version of this experiment (Richler et al., 2011). In the complete design, holistic processing is indexed by an interdependence of congruency and alignment (Richler et al., 2011): performance is better in congruent than in incongruent trials (i.e. congruency effect). Misalignment reduces the congruency effect, as it disrupts holistic processing. Following McKone and colleagues' advice (McKone et al., 2013) we tested the composite face effect in upright and inverted conditions. The inverse condition, like misalignment, also disrupts holistic processing. Therefore, inversion in interdependence with congruency also measures holistic processing: the congruency effect is larger for upright than inverted trials.

Task

In each trial, participants had to indicate whether the two face halves comprising the eyes were the same or not. Participants responded during the inter-trial interval of 2 s by pressing the relevant keys on the keyboard. All participants were able to answer within the inter-trial interval for all trials, i.e. there were no trials without response. No feedback was given. After every 20 trials and also between blocks participants were able to take a self-paced break. Before testing, there were 10 training trials for each of the four different blocks. Blocks were trained in the same order, as they would appear during the actual testing.

Results

For each participant we calculated the d'-scores as Z(hits = accuracy in same trials) − Z(false alarms = 1 - accuracy in different trials). The congruency effect was calculated by subtracting d'-scores of incongruent from congruent conditions. Figure 8 depicts the mean congruency effects per group. In the upright condition controls obtained a mean congruency effect of 1.33 (SD = 1.27) for aligned, and 0.1 (SD = 0.8) for misaligned trials, while prosopagnosics obtained a mean congruency effect of 0.65 (SD = 0.61) for aligned, and 0.31 (SD 0.48) for misaligned trials. In the inverted condition controls obtained a mean congruency effect of 0.36 (SD = 0.70) for aligned, and 0.14 (SD = 0.52) for misaligned trials, while prosopagnosics obtained a mean congruency effect of -0.11 (SD = 0.46) for aligned, and 0.27 (SD = 0.45) for misaligned trials.

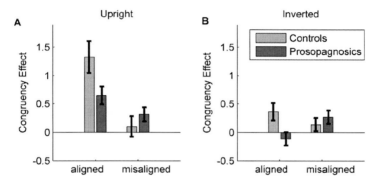

Figure 8: (A) Mean size of the congruency effect in the upright condition for controls and prosopagnosics. Error bars: SEM. (B) Mean size of the congruency effect in the inverted condition for controls and prosopagnosics. Error bars: SEM.Example stimuli of the composite face task.

As misalignment and inversion are both control conditions for the measurement of holistic processing, we consider these two factors separately. First, we looked at the congruency effect for the upright condition only, using misalignment as control condition. A two-way repeated measures ANOVA on participant group (prosopagnosics, controls) and alignment (aligned, misaligned) was conducted. The congruency effect was larger for the aligned than the misaligned conditions ($F(1, 35) = 23.54$, $p < .001$, $\eta^2 = .40$) and there was no significant difference between participant groups ($F(1, 35) = 0.93$, $p = .34$, $\eta^2 = .03$). The interaction between alignment and participant group was significant, indicating that the congruency effect of controls was more affected by misalignment than the congruency effect of prosopagnosics ($F(1, 35) = 7.71$, $p = .009$, $\eta^2 = .18$). A post hoc analysis for prosopagnosics only revealed that their congruency effect was significantly smaller for the misaligned than aligned condition (one-way ANOVA: $F(1, 15) = 12.90$, $p = .003$, $\eta^2 = .34$). This indicates that prosopagnosics still exhibit holistic processing.

Second, we looked at the congruency effect for the upright-aligned versus the inverted-aligned conditions only, using inversion as control condition. A two-way repeated measures ANOVA for the aligned condition on orientation (upright, inverted) and participant group (prosopagnosics, controls) was conducted. As expected, the congruency effect was larger for upright than inverted conditions ($F(1, 35) = 16.0$, $p < .001$, $\eta^2 = .31$) and controls showed overall a larger congruency effect than prosopagnosics ($F(1, 35) = 10.11$, $p = .003$, $\eta^2 = .22$). The interaction between orientation and group was non-significant, indicating that the inversion factor did not affect prosopagnosics and controls differently ($F(1, 35) = 0.23$, $p = .64$, $\eta^2 = .01$).

Additionally, we investigated more closely the negative congruency effect observed for prosopagnosics in the inverted-aligned condition (see Figure 8.B). The congruency effect was significantly smaller for aligned than misaligned trials in the inverted condition for prosopagnosics ($F(1, 31) = 7.29$, $p = .016$, $\eta^2 = .16$). This was not the case for controls, who showed no difference in congruency effects ($F(1, 41) = 1.27$, $p = .27$, $\eta^2 = .03$). This negative effect for prosopagnosics is called the "inversion superiority effect" in previous studies (e.g. de Gelder et al., 1998; Behrmann and Avidan, 2005).

Discussion

The congruency effect in interdependence with alignment or orientation serves as a measure of holistic processing. For the upright condition, using misalignment as control condition, we found that controls showed a larger difference in congruency effect for aligned versus misaligned trials compared to prosopagnosics. These results suggest that holistic processing is impaired or utilized to a smaller extent by prosopagnosics in this task. This replicates the results of previous reports of decreased holistic processing for prosopagnosics compared to controls (Avidan et al., 2011; Palermo et al., 2011)[2]. However, as the difference in congruency effect for aligned versus misaligned trials is still significant for prosopagnosics, this indicates that their holistic processing ability is still present, yet impaired.

---

[2] Note, though, that in those studies only the partial design was used and only with upright faces.

When using the inversion effect as control (upright-aligned versus inverted-aligned conditions only), we did not find a significant difference between groups in interdependence of congruency effect and inversion, as indicated by the non-significant interaction. Thus, we were not able to find differences in holistic processing between groups, which is contrary to the expectations given by our design. Furthermore, our results imply that prosopagnosics show more holistic processing for misaligned face halves than aligned halves when seen inverted, as indicated by the significant interdependence between congruency and alignment in the inverted condition for prosopagnosics. Similar "inversion superiority effects" for prosopagnosics have been described before (Susilo et al., 2010; Farah et al., 1995; Behrmann et al., 2005). However, what exactly happens when prosopagnosics process inverted faces is currently not well understood. For this reason, the results in the inverted condition should not be seen as indicator of holistic processing for prosopagnosics. We would thus argue that the advice to run the composite face effect in upright and inverted conditions (McKone et al., 2013) is not suitable for testing prosopagnosics participants. Nevertheless, the fact that the controls showed the expected pattern in the inverted condition (small congruency effects with no difference between alignment conditions), supports the general validity of the used method.

In sum, if we concentrate on the upright condition that can be clearly interpreted, prosopagnosics, compared to controls, show a smaller difference between the congruency effects obtained in the aligned and misaligned condition. This indicates that holistic processing (as indexed by the interdependence of congruency and alignment) is impaired for prosopagnosics. To get a closer and more detailed look at the impairment of holistic processing in prosopagnosia, we tested what type of facial information retrieval is impaired in the study described next.

4.5.   Featural and configural sensitivity task

Motivation

The previous test revealed that holistic processing is impaired in prosopagnosics. Several studies found that configural and featural processing, which contribute to holistic processing, are also impaired in prosopagnosics. One study used blurred (disrupted featural information with intact configural information) and scrambled (disrupted configural information with intact featural information) face stimuli and found significantly decreased recognition scores in both cases for prosopagnosics compared to controls (Lobmaier et al., 2010). Similarly, a further study found a featural and configural disadvantage for discriminating faces differing either in features or configuration (Yovel and Duchaine, 2006). In contrast, another study reported that only a minority of the tested prosopagnosics showed a reduced sensitivity on a same-different task with faces differing either in features or configuration (Le Grand et al., 2006). The face stimuli used in these studies either contained unnatural modifications (scrambling and blurring, Lobmaier et al., 2010), had configural modifications beyond natural limits (Yovel and Duchaine (2006), as discussed in Maurer et al. (2007)), or contained extra clues like makeup (Le Grand et al., 2006). In view of these conflicting results, which might have been induced by the use of unnatural or clue-containing stimuli, we used a stimulus set of natural looking faces, with parametric differences in features and configuration, to investigate the sensitivity to featural and configural facial information of prosopagnosics compared to controls.

Stimulus creation and task have been described in detail elsewhere (Esins et al., 2014b). Therefore, we will give only a short description here.

Stimuli

We manipulated male faces from the Max Planck 3D face database to create eight face sets. For creating each set, different faces were used. In each set, the faces differed in features (eyes, nose, and mouth) or their configuration, but they shared the same skin texture and outer shape[3] (see Figure 9). Changes in features and in configuration were implemented parametrically resulting in five similarity levels from 100 % (identical faces) to 0 % (maximal difference within each set) between the faces. The central faces of both dimensions (features and configuration) are identical for each set.

The stimuli had a visual angle of 5.7 degrees horizontally and 8.6 degrees vertically. To prevent pixel matching the faces were presented at different random positions on the screen within a viewing angle of 7.6 degrees horizontally and 10.5 degrees vertically.

---

[3] In a previous study (Esins et al., 2011), the natural appearance of these faces has been controlled.

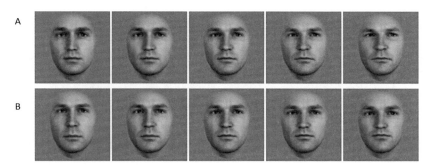

Figure 9: Faces of one set, (A) differing in features while their configuration stays the same and (B) differing in configuration while their features stay the same. Skin texture and outer face shape were kept constant within each set. The middle faces of both rows are the same.

Task

Participants rated the perceived pair-wise similarity of the faces within each set on a Likert scale from 1 (very little similarity) to 7 (high similarity / identical). They were advised to use the whole range of ratings during the experiment. In each trial, the first face was displayed for 2 s, followed by a pixelated face mask for 0.8 s, and then the second face for another 2 s. Afterwards the Likert scale was displayed and participants marked their rating by moving a slider on the scale via the arrow keys and confirmed their choice by pressing the relevant key on the keyboard. The start position of the slider was randomized. The next trial started as soon as the rating was confirmed. There were no time restrictions, but participants were told to answer without too long considerations. After every 20 trials participants could make a self-paced break.

The faces of each set were compared with each other and with themselves. We were only interested in trials comparing faces manipulated along the same dimension (see Figure 9.A for features and Figure 9.B for configuration). Filler trials in which faces differed in features and configuration were displayed during the test to avoid participants realizing the nature of the stimuli. These filler trials were omitted from the analysis. For each participant the order of trials was randomized within and across sets.

Results

For each participant we calculated the mean ratings for each of the five similarity levels across all sets, but separately for each change type (featural, configural). Similarity ratings were close to seven (high similarity) for identical faces and dropped with decreasing similarity. For each participant we fitted a linear regression to the mean similarity ratings, again separately for featural and for configural ratings. The similarity levels served as predictors. The steepness of the slopes was then used as measure of sensitivity: the steeper the slopes, the stronger the participant perceived the configural or featural changes.

Figure 10 depicts the mean sensitivity scores per group. Controls obtained a mean sensitivity score of 1.23 (SD = 0.23) for featural, and 0.70 (SD = 0.15) for configural changes. Prosopagnosics obtained a mean sensitivity score of 1.09 (SD = 0.20) for featural, and 0.51 (SD = 0.23) for configural changes. We analyzed the sensitivity scores with a two-way repeated-measures ANOVA with the factors change type (features, configuration) and participant group (prosopagnosics, controls). Participants exhibited a higher sensitivity towards featural than configural changes ($F(1, 35) = 172.76$, $p < .001$, $\eta^2 = .83$), and prosopagnosics showed an overall reduced sensitivity compared to controls ($F(1, 35) = 9.34$, $p = .004$, $\eta^2 = .21$). The interaction between change type and participant group was non-significant ($F(1, 35) = 0.41$, $p = .53$, $\eta^2 = .01$). For the analysis of simple effects we used a one-way ANOVA: For changes in features the controls showed only a marginally significantly larger sensitivity ($F(1, 36) = 3.61$, $p = .066$, $\eta^2 = .09$) whereas for changes in configuration controls showed a significantly larger sensitivity than prosopagnosics ($F(1, 36) = 8.89$, $p = .005$, $\eta^2 = .20$).

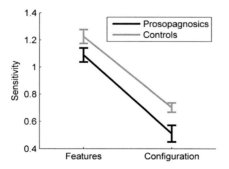

Figure 10: Mean sensitivity to features and configuration for controls and prosopagnosics. Error bars: SEM.

Discussion

Our results show that the sensitivity to configural information is significantly impaired for prosopagnosics compared to controls. We also find a trend that the sensitivity to featural information is impaired in prosopagnosics (p = .066). These results bridge the gap between findings of previous studies. On the group-wise level, prosopagnosics were found to have a significantly lower accuracy than controls for faces modified in configuration, but not for faces modified in features (calculated from Table 4 in Le Grand et al., 2006). This result was confirmed by Yovel and Duchaine (2006) using the same stimuli as Le Grand and colleagues for their prosopagnosic participants. However, Yovel and Duchaine also tested their participants with differently modified faces (stronger modifications in configuration) and found a reduced sensitivity to both information types for prosopagnosics (Yovel and Duchaine, 2006). In a study of Lobmaier and colleagues (2010) all prosopagnosics performed more than two standard deviations below controls' mean performance on configural (blurred) trials, while only 4 of the 6 prosopagnosics performed more than two standard deviations below controls' mean performance on featural (scrambled) trials. Nevertheless, prosopagnosics obtained significantly lower recognition scores for both, featural and configural information (Lobmaier et al., 2010). Taken together, the results of previous studies and our results show that the retrieval of configural information is impaired in prosopagnosics. For the sensitivity to features, our results bridge the non-significant findings of one study (Le Grand et al., 2006) and the significant results of other studies (Yovel and Duchaine, 2006; Lobmaier et al., 2010) Therefore, we conclude that the retrieval of featural information might be impaired for prosopagnosics, although to a lesser degree than the retrieval of configural information.

4.6. Gender recognition task

Motivation

Most prosopagnosics self-report normal recognition of gender (Grüter et al., 2008) which is also reflected by the results of behavioral studies (DeGutis et al., 2012; Chatterjee and Nakayama, 2012; Le Grand et al., 2006). Nevertheless there are some single-case studies which report prosopagnosics' gender recognition to be impaired (De Haan and Campbell, 1991; Ariel and Sadeh, 1996; Jones and Tranel, 2001; Duchaine et al., 2006). In view of those equivocal reports, we aimed at clarifying this open issue.

Stimuli

We used 80 faces (40 male) from our Max Planck 3D face database. As visible in Figure 11 the faces contained no additional hints about the gender, like hair, beard, or makeup. The stimuli had a visual angle of 3.8 degrees horizontally and 5.7 degrees vertically.

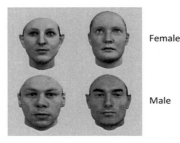

Female

Male

Figure 11: Example of female and male faces used as stimuli for the gender recognition task.

Task

Participants had to judge the gender of each face. The faces were shown one at a time and stayed on screen until a response was given by pressing the relevant keys on the keyboard. The next image appeared as soon as a response was entered. The order of trials was randomized. No feedback was given. Participants were instructed to answer as correctly and as quickly as possible.

Results

For each participant the percent correct accuracy was calculated. Figure 12 depicts the mean scores per group. Controls achieved a very high mean accuracy of 91.5 % (SD = 4.8), while prosopagnosics scored very well too at 84.4 % (SD = 5.9). Nevertheless, an one-way ANOVA revealed that prosopagnosics performed significantly worse than controls ($F(1, 36) = 16.62, p < .001, \eta^2 = .32$).

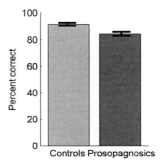

Figure 12: Mean percent correctly classified faces in the gender recognition task for controls and prosopagnosics. Error bars: SEM

Discussion

Prosopagnosics exhibited a significantly lower gender recognition ability compared to controls. This differs from the self-reports of prosopagnosics (Grüter et al., 2008), and also from behavioral tests in several studies (DeGutis et al., 2012; Chatterjee and Nakayama, 2012; Le Grand et al., 2006). However, there are some prosopagnosic single case studies which report impairments of gender recognition (De Haan and Campbell, 1991; Ariel and Sadeh, 1996; Jones and Tranel, 2001; Duchaine et al., 2006). In our test, we observed high performance for the control group and comparatively high performance for the prosopagnosics. We argue that first, prosopagnosics suffer from only a slight impairment of gender recognition and second, that this impairment may be easily compensable in daily life by using cues like body shape, hairdo, makeup, voice, etc. Our conclusion that the gender recognition impairment is only slight and easily compensable, is supported by the fact that controls and prosopagnosics achieved ceiling performance in gender recognition tests in several further studies (Dobel et al., 2007; Gruber et al., 2011; Lobmaier et al., 2010). In our study, we used well controlled stimuli derived from real faces. It is possible that this type of stimuli and our large sample size helped to reveal the gender recognition deficit in prosopagnosics. Along this line, another study which also used faces of the same Max Planck 3D face database, showed impaired same-or-different recognition performance for faces differing in gender for their prosopagnosic participants (Behrmann et al., 2005).

4.7.   Facial motion advantage

Motivation

Most studies testing holistic face recognition abilities of prosopagnosics use only static face stimuli. Furthermore, those face images often are identical for training and testing. Such tasks do not reflect the everyday challenges encountered by prosopagnosics, as people move, speak and might alter their look on a day-to-day basis. On the one hand, the different looks of people likely complicate recognition for prosopagnosics, maybe even more than for controls, who do not rely on these non-facial attributes for recognition. On the other hand, the additional dynamic information might give additional cues for prosopagnosics, thus facilitating recognition ('motion advantage'), potentially also even more than for controls, who, again, do not rely on this additional information for recognition. Therefore, we wanted to investigate the influence of look and motion on face recognition for prosopagnosics compared to controls, by using dynamic and changeable stimuli. We also included static and non-changing (identical) stimuli as control conditions.

Stimuli

The stimuli used in this test have been created and kindly provided by O'Toole and colleagues (O'Toole et al., 2005). Recordings of 72 actresses speaking into the camera, expressing natural rigid and non-rigid movements served as 'dynamic stimuli' (Figure 13.A). The 'static stimuli' displayed five random frames from the original recordings, shown for 1 s each and separated by a black screen for 0.2 s (Figure 13.B). There were two recordings of each actress. In the second recordings, the actresses had a different hairdo, makeup, or accessories (see Figure 14). These 'changed-look' recordings also were prepared as dynamic or static stimuli. Actresses were shown only in one of both conditions (static or dynamic) during the task.

All stimuli presented the faces for 5 s and were mute. Each actress was placed in front of a grey background and her clothing was covered. The stimuli had a visual angle of 16.6 degrees horizontally and 12.4 degrees vertically.

A Dynamic stimuli

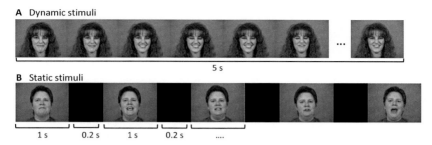

5 s

B Static stimuli

1 s   0.2 s   1 s   0.2 s   ....

Figure 13: Example of stimuli. (A) dynamic stimuli: recordings of persons speaking. (B) static stimuli: five random frames extracted from the original recordings are shown one after the other. Each frame is shown for 1 s with a black frame for 0.2 s between frames.

Figure 14: Example of changes in looks.

Task

The experiment consisted of a dynamic and a static block. In the dynamic block, dynamic stimuli were used for learning and testing, and static stimuli in the static block. Block order was counterbalanced between participants. In each block participants first learned 18 target identities and then performed an old-new recognition test on the 18 target identities and 18 distractor identities. Half of the targets looked identical during learning and at test, while the other targets were displayed with a change in their appearance at test. Participants were informed that the look might change between learning and testing. The order of trials was randomized during learning and testing. Which target actresses were tested in the identical or changed look condition was counterbalanced across participants. During learning and testing participants could see each stimulus only once. During testing, the stimuli were presented for the length of their duration or until key press, whichever came first. The next stimulus started as soon as an answer was entered by pressing the relevant keys on the keyboard. No feedback was given. Between blocks participants were able to take a self-paced break.

Results

We calculated d'-scores for each participant. Figure 15 depicts the mean scores per group in the identical and the changed look condition. For the identical look condition, controls achieved a mean d'-score of 2.79 (SD = 0.53) for dynamic stimuli and 2.25 (SD = 0.65) for static stimuli. Prosopagnosics achieved a mean d'-score of 1.87 (SD = 0.74) for dynamic stimuli and 1.85 (SD = 0.61) for static stimuli. For the changed look condition, controls achieved a mean d'-score of 1.72 (SD = 0.71) for dynamic stimuli and 1.40 (SD = 0.71) for static stimuli. Prosopagnosics achieved a mean d'-score of 1.09 (SD = 0.76) for dynamic stimuli and 1.19 (SD = 0.48).

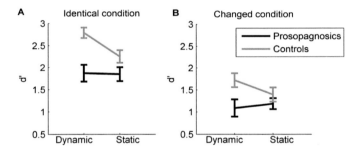

Figure 15: (A) Mean d'-scores for the identical look condition for controls and prosopagnosics. Error bars: SEM. (B) Mean d'-scores for the changed look condition for controls and prosopagnosics. Error bars: SEM.

We ran a two-way repeated measures ANOVA for the factors look (identical, changed) versus participant group (prosopagnosics, controls). We found a significantly better performance in the identical look condition ($F(1, 35) = 117.32$, $p < .001$, $\eta^2 = .77$), and a significantly better performance for controls ($F(1, 35) = 15.46$, $p < .001$, $\eta^2 = .31$). The interaction of look and participant group was non-significant ($F(1, 35) = 2.42$, $p = .13$, $\eta^2 = .07$), indicating that there was no difference between controls and prosopagnosics in how the look of the identities influenced their performance.

We also conducted two-way repeated measures ANOVAs on stimulus type (dynamic, static) and participant group (prosopagnosic, control) for the identical and for the changed look condition separately. For the identical look, we found a statistically better performance for dynamic than static stimuli ($F(1, 35) = 4.99$, $p = .032$, $\eta^2 = .13$), and a statistically better performance for controls than prosopagnosics ($F(1, 35) = 15.62$, $p < .001$, $\eta^2 = .31$). Interaction between stimulus type and participant group was significant, too ($F(1, 35) = 4.14$, $p = .0496$, $\eta^2 = .11$). For the changed look condition there was no performance difference for static versus dynamic stimuli ($F(1,35) = 0.44$, $p = .51$, $\eta^2 = .01$). We found a better performance for controls than prosopagnosics ($F(1,35) = 8.03$, $p = .008$, $\eta^2 = .19$). The interaction between stimulus type and participant group was non-significant ($F(1, 35) = 1.54$, $p = .22$, $\eta^2 = .04$).

Post hoc analysis of simple effects of stimulus type (dynamic, static) revealed that controls performed better for dynamic than static stimuli in the identical look condition (one-way ANOVA, $F(1, 41) = 8.65$, $p = .005$, $\eta^2 = .18$). For all other post hoc tests the differences were non-significant (one-way ANOVAs, $Fs < 2.21$, $ps > .14$, $\eta^2s < .06$). Therefore, controls recognized identical looking identities better in the dynamic than in the static condition, while prosopagnosics did not show a difference in performance for dynamic compared to static stimuli.

Discussion

The first finding of this test is that controls and prosopagnosics showed a similar drop in recognition performance when the look of a person changes compared to when the look stays the same. Therefore, we could not find evidence that prosopagnosics are more affected than controls when unfamiliar faces change look between learning and testing.

The second finding reveals that controls, but not prosopagnosics, showed a facial motion advantage when trained and tested with identical looking stimuli. These results are in line with a study which also tested prosopagnosics on dynamic and static face stimuli with an old-new recognition task with faces either upright or inverted (Longmore and Tree, 2013). In the upright condition, prosopagnosics showed no significant difference in performance for dynamic and static stimuli, while controls were better for the dynamic stimuli. Longmore and colleagues' interpretation was, that the task was too difficult for the prosopagnosic participants, making it impossible to detect a facial motion advantage for prosopagnosics (mean accuracy rates of the prosopagnosics were about 60 % for both static and dynamic upright stimuli, with chance level being 50%). We find similarly no motion advantage for prosopagnosics in our identical condition. However, our task did not seem to be too difficult: Prosopagnosics showed mean d'-scores between 1.0 and 2.0, with d' = 0 corresponding to chance level. Therefore, we suggest that our results provide a valid measure of the absence of a motion advantage for prosopagnosics, as the explanation of a too difficult task is not given in our case.

In the changed condition, our results show a similar, but non-significant trend: controls exhibit a motion advantage while prosopagnosics do not. In an earlier report of that study, we had found a significant motion advantage for controls in the changed condition (Esins et al., 2014a). At that time, we had analyzed recognition performance of the same 16 prosopagnosic participants and 16 of the 21 controls reported here, matched to the prosopagnosics in age and gender as closely as possible. Therefore, we suggest, that a larger sample size is needed to verify the robustness of this finding.

Further support of the lack of a motion advantage for prosopagnosics reported here is given by a study finding impaired biological motion perception for face, but not whole-body stimuli for congenital prosopagnosic participants (Lange et al., 2009).

Taken together, these results hint to a lack of a motion advantage for prosopagnosics. This could be explained by a neurophysiological dysfunction in prosopagnosia that affects not only the ventral temporal face-processing regions but also the lateral temporal facial motion-processing regions, in particular the superior temporal sulcus (Hoffman and Haxby, 2000), a core region of the face processing network (Ishai et al., 2005). The right posterior STS was found to have a significantly reduced connectivity with the other core regions of the face processing network in prosopagnosics (Avidan et al., 2013). This finding is in accordance with the fact that we find no motion advantage for prosopagnosics observed in our study.

Table 2: Prosopagnosics' behavioral data for the tests: z-scores.

	4.1. CFMT	4.2. Cars		4.3. Surprise recognition		4.4. Composite face task				4.5. Featural and configural sensitivity task		4.6. Gender	4.7. Facial motion advantage			
		CCMT	Car expert-ise	sur-prise	control	up-algn	up-misalgn	inv-algn	inv-misalgn	Feat-ures	Config-uration		id-dyn	id-stat	ch-dyn	ch-stat
1	-1.96*	1.43	0.26	0.68	-1.77	-0.49	0.48	-1.55	-1.06	-0.66	-0.86	-2.93*	-0.65	-0.10	-0.51	-0.41
2	-3.28*	-1.38	-2.07*	-1.39	-1.12	-0.74	0.17	-0.90	0.83	0.41	1.50	-2.93*	-0.81	-0.35	-1.55	0.89
3	-3.14*	-1.27	-2.46*	-1.42	-0.87	-1.43	0.39	-1.31	-0.64	-1.29	0.81	-2.41*	-0.84	-1.71	-1.97*	-0.35
4	-2.26*	1.54	1.42	-2.12*	-0.13	-1.54	-0.70	-0.09	-1.66	-1.29	-1.75	-2.93*	-3.01*	-0.41	-3.54*	-0.28
5	-1.81	0.13	-2.07*	-1.00	-3.09*	-0.23	-0.13	-0.59	0.40	-1.16	0.57	-0.05	-4.19*	-1.32	-1.15	0.47
6	-2.99*	-0.62	-1.68	-1.72	-1.82	-0.46	0.88	0.38	0.72	-0.77	-0.41	-1.36	-3.01*	-1.77	-1.15	-1.21
7	-3.73*	-1.81	-0.91	-1.24	-1.77	-0.51	-1.07	-0.65	0.02	-0.80	-0.49	-0.57	-2.43*	-3.04*	-0.95	-1.08
8	-3.14*	-0.19	-1.29	-0.61	-1.22	0.43	0.70	-1.15	0.59	-1.54	-2.65*	0.74	-1.52	-0.81	-1.15	-1.21
9	-2.84*	0.35	0.26	-0.80	-0.49	-0.74	0.64	-0.97	1.10	1.51	-3.91*	-1.10	-2.96*	-0.68	-1.82	-0.53
10	-2.70*	-0.73	-0.52	-1.60	0.81	-0.48	0.24	-0.04	0.92	0.05	-0.98	-0.05	0.75	0.16	0.90	0.48
11	-3.58*	1.00	0.26	-0.26	-0.70	-0.15	0.55	-1.66	0.16	-0.47	-1.14	-1.88	0.05	0.48	-0.10	-0.80
12	-2.99*	-0.19	-2.07*	-1.42	-2.91*	-0.01	0.84	-1.46	1.02	-0.47	-2.32*	-2.15*	-2.52*	0.29	0.02	-0.44
13	-1.96*	-0.51	-0.91	-1.61	-1.06	-0.73	0.25	-0.72	-0.81	-1.37	-2.89*	-2.93*	-3.07*	-0.80	-1.26	-0.33
14	-3.58*	0.56	1.42	-0.80	-1.48	-0.22	0.29	-0.37	0.81	-1.70	-3.46*	-1.10	-1.99*	0.29	0.02	-0.05
15	-2.11*	-1.05	-1.68	-1.60	-1.64	-0.60	-0.50	0.39	1.36	0.71	-1.10	-0.05	-1.52	-0.41	0.37	-0.69
16	-2.40*	-0.19	-0.52	-2.18*	-1.41	-0.70	1.19	-0.18	0.31	-0.77	-0.74	-2.15*	0.05	0.16	-0.38	0.89
Mean	-2.78	-0.18	-0.79	-1.19	-1.29	-0.54	0.26	-0.68	0.25	-0.60	-1.24	-1.49	-1.73	-0.63	-0.89	-0.29
SD	0.63	0.98	1.23	0.72	0.96	0.48	0.61	0.66	0.87	0.88	1.53	1.23	1.40	0.95	1.07	0.68

Z-scores were calculated for each tests based on mean and standard deviation of controls' results obtained in that test. Z-scores of less than -1.96 are marked with an asterisk.

Table 3: Controls' behavioral data for the tests: z-scores.

	4.1. CFMT	4.2. Cars		4.3. Surprise recognition		4.4. Composite face task				4.5. Featural and configural sensitivity task		4.6. Gender	4.7. Facial motion advantage			
		CCMT	Car expert-ise	sur-prise	control	up-algn	up-misalgn	inv-algn	inv-misalgn	Feat-ures	Config-uration		id-dyn	id-stat	ch-dyn	ch-stat
1	0.39	1.21	-0.91	-0.29	0.45	-0.01	0.27	0.59	-0.59	0.58	0.28	-0.84	0.05	0.16	0.37	-0.69
2	-0.05	0.67	-0.13	-0.59	-0.49	2.75	0.21	0.39	-1.54	-0.80	-0.21	-2.41*	-1.14	-2.52*	0.13	-2.36*
3	1.86	1.21	1.81	1.48	1.17	0.79	-0.13	0.70	-0.26	0.58	1.05	-1.10	-2.15*	0.87	-0.10	2.00
4	0.83	-0.51	-0.91	1.86	1.17	0.06	0.54	-0.52	0.58	-0.83	0.61	-0.84	-0.65	-1.39	-0.10	-0.53
5	1.57	0.67	1.81	0.14	-0.68	1.39	0.39	-0.31	0.77	-1.07	0.00	-0.05	0.75	0.87	0.90	-0.05
6	0.39	0.24	-0.52	1.11	-1.41	-0.07	0.21	1.15	0.16	0.71	-0.65	0.21	1.36	0.87	-0.33	-0.44
7	0.25	-1.70	-0.13	-0.80	-1.82	-1.00	-0.69	-0.38	0.06	1.18	1.74	0.47	-0.43	-0.81	-0.87	-0.80
8	1.57	0.03	-0.91	1.05	1.17	-0.82	-1.58	0.90	0.80	0.38	-0.09	-0.57	0.75	-0.81	1.54	1.12
9	-1.52	-0.19	-0.13	-1.06	-0.70	-1.48	-0.30	2.58	0.21	0.00	1.13	0.47	1.36	-1.32	0.08	0.47
10	-0.93	-1.05	0.65	-1.24	0.81	-0.35	1.84	0.73	-0.26	0.11	-1.14	1.00	0.05	-0.41	-0.10	-0.85
11	-1.52	-1.70	-0.52	-1.63	1.17	-0.17	-0.05	-0.20	0.81	1.51	1.01	-0.57	0.05	0.29	-0.51	0.83
12	-0.34	-2.02*	-2.46*	0.38	-0.34	-0.03	0.27	-2.18*	0.20	-0.41	-0.29	0.74	0.05	1.45	0.37	-0.33
13	-1.23	0.24	0.65	0.14	-1.41	-0.40	0.02	-0.52	1.98	-1.70	-0.86	0.47	-0.81	-0.80	-1.26	0.08
14	-0.64	-0.84	-0.91	0.38	0.45	0.14	-1.18	-1.73	-0.19	1.29	0.12	1.00	0.75	0.29	0.90	-0.44
15	0.54	1.21	0.65	0.11	-0.34	-0.68	-1.46	0.04	-0.69	-1.59	-0.37	0.74	1.36	-0.11	2.52	-0.53
16	-0.93	0.46	-0.13	0.14	0.03	1.78	2.56	-0.13	0.87	0.77	1.66	0.47	-1.52	0.29	-0.90	-0.05
17	-0.05	-0.08	1.42	-1.60	-0.26	-0.17	-0.13	-0.14	-2.51*	-0.25	0.08	0.21	0.05	0.87	-0.38	0.48
18	-0.78	0.46	-0.13	-1.04	1.17	-0.83	-1.54	-0.20	-0.59	0.33	-1.43	1.52	0.75	0.87	-0.79	-0.05
19	0.39	0.78	0.26	1.05	1.17	-1.03	0.56	-0.07	-1.33	-0.80	-2.12*	-0.05	-1.44	0.87	-2.02*	2.00
20	0.83	1.21	0.26	0.68	0.10	0.43	0.24	-0.99	0.80	1.40	0.40	1.00	0.75	0.87	-0.38	0.83
21	-0.64	-0.30	0.26	-0.26	-1.40	-0.28	-0.05	0.28	0.73	-1.37	-0.94	-1.88	0.05	-0.41	0.90	-0.69
Mean	0.00	0.00	0.00	0.00	0.00	0.00	0.00	0.00	0.00	0.00	0.00	0.00	0.00	0.00	0.00	0.00
SD	1.00	1.00	1.00	1.00	1.00	1.00	1.00	1.00	1.00	1.00	1.00	1.00	1.00	1.00	1.00	1.00

Z-scores were calculated for each tests based on mean and standard deviation of controls' results obtained in that test. Z-scores of less than -1.96 are marked with an asterisk.

64

Table 4: Reliability coefficients, statistical significance of difference and ratio between groups' reliability coefficients for each test.

Tests	Parts	Prosopagnosics		Controls		p		Ratio	
		Cronbach	adjusted split-half	Cronbach	adjusted split-half	Cronbach	split-half	Cronbach	adjusted split-half
4.1. CFMT		0.26	0.31	0.82	0.83	0.01*	0.05	3.15	2.68
4.2. Cars	CCMT	0.87	0.88	0.88	0.89	0.83	0.89	1.01	1.01
	Car-expertise	0.69	0.72	0.56	0.58	0.49	0.58	0.81	0.81
4.3. Surprise recognition	surprise	-	0.30	-	0.66	-	0.32	-	2.20
	control	-	0.65	-	0.76	-	0.62	-	1.17
4.4. Composite face task	up-algn	0.85	0.86	0.94	0.95	0.04*	0.16	1.11	1.10
	up-misalgn	0.86	0.87	0.92	0.93	0.27	0.42	1.07	1.07
	inv-algn	0.86	0.87	0.92	0.92	0.28	0.43	1.07	1.06
	inv-misalgn	0.83	0.84	0.91	0.92	0.24	0.40	1.10	1.10
4.5. Featural and configural sensitivity task	Features	0.97	0.97	0.94	0.94	0.18	0.35	0.97	0.97
	Configuration	0.97	0.97	0.96	0.96	0.65	0.74	0.99	0.99
4.6. Gender recognition		0.67	0.70	0.69	0.72	0.88	0.91	1.03	1.03
4.7. Facial motion advantage	Dynamic	-	0.52	-	0.44	-	0.83	-	0.85
	Static	-	0.17	-	0.50	-	0.48	-	2.94

Reliability coefficients were calculated with bootstrapped, Spearman-Brown adjusted split-half method and, where possible, with Cronbach's alpha. Statistical significance of difference between groups' reliability coefficients was calculated as difference between correlations for unadjusted split-half reliability coefficients, and with the Bonett formula for Cronbach's alpha (Bonett, 2003). Statistical significant p-values are marked with an asterisk. The ratio between prosopagnosics and controls reliability coefficients was calculated by dividing controls reliability coefficients by prosopagnosics reliability coefficients.

5.   Reliabilities

As our study included newly developed tests, we assessed their reliability by calculating the reliability scores for each participant group separately. For consistency reasons we did that also for the established tests we used, especially as this has never been calculated for purely prosopagnosic populations. The reliability is an indicator of the consistency quality of a test. We calculated the internal consistency reliability with Cronbach's alpha. This reliability measure indicates if trials measuring the same face recognition mechanisms give similar results; thus indicating a good test consistency. This was done for all tests except for the surprise recognition (test 4.3), and the facial motion advantage (test 4.7). For both tests, Cronbach's alpha could not be calculated, because participants saw different stimuli, depending on assignment. For this reason, we also calculated the reliability coefficient with a bootstrapped split-half method with Spearman-Brown correction, which is mathematically related to Cronbach's alpha and can be applied to all our tests. Both coefficients gave similar results, validating the calculation (see Table 4). Additionally, we calculated the statistical difference between controls and prosopagnosics for either coefficient; the p-values are given in Table 4 as well. In the further course of this manuscript, we refer to Cronbach's alpha as reliability coefficient, if both coefficients (Cronbach's alpha, split-half estimate) are given. Tests 4.2, 4.3, 4.4, 4.5, and 4.7 comprised of several parts testing different aspects of face recognition. Therefore, we calculated Cronbach's alpha and split-half estimate for the test parts separately.

For Cronbach's alpha, coefficients of more than .7 indicate acceptable to excellent reliability (Nunnally, 1978; Lance et al., 2006). For controls, reliability coefficients close to .7 and higher (and mostly larger than .9) were reached in all tests parts but four (see Table 4). For prosopagnosics, most reliability coefficients were similar to those obtained by the controls and deviated by less than 20 % (i.e. the ratio of reliability coefficients between groups was between 0.8 and 1.2. See Table 4). However, for four tests parts prosopagnosics reliability coefficients conspicuously deviated from controls' coefficients. For three test parts, controls exhibited more than two to three times higher reliability coefficients (i.e. for these three test parts the ratio of reliability coefficients was larger than 2.2, indicating a deviation of 120 % and more between controls and prosopagnosics. See Table 4). The tests parts are (4.1) the CFMT, (4.3) surprise condition of the surprise recognition, and (4.7) static condition of the facial motion advantage. The difference of reliability coefficients between groups reached significance for the CFMT, but not for the other two test parts. The fourth test part with conspicuous deviation between controls and prosopagnosics is the upright-aligned condition for test (4.4) the composite face task. Here as well the difference of reliability coefficients between controls and prosopagnosics was significant.

A literature search for experimental reliability coefficients for the CFMT found only studies reporting Cronbach's alpha for control participants: Cronbach's alpha = .83 (Herzmann et al., 2008), Cronbach's alpha = .90 (Wilmer et al., 2010), and Cronbach's alpha = .89 (Bowles et al., 2009). We were not able to find a study reporting reliability for the CFMT for purely prosopagnosic participant groups. Therefore, we report here for the first time this surprising result.

Importantly, all four test parts, for which prosopagnosics showed a conspicuous deviation of their reliability coefficients compared to controls, test for holistic recognition of whole, static faces: namely (4.1) CFMT, (4.3) surprise condition of the Surprise recognition, (4.7) static condition of the Facial motion advantage, and (4.7) upright aligned condition of the Composite face task. The other test parts do not test holistic face recognition but rather face classification, featural and configural processing, face parts comparison, object recognition, or deal with moving faces. The fact that there is no reduced reliability for recognition of dynamic faces in test (4.7) Facial motion advantage could have several causes. One possible explanation is that other than holistic processes are activated when recognizing dynamic faces, which allow for performance of prosopagnosics to be more consistent. This hypothesis is supported by a study finding that non-rigid face motion promotes feature-based processing rather than holistic processing in their laboratory condition (Xiao et al., 2013).

For test (4.1) CFMT, we conducted further analyses to investigate the reduced reliability of prosopagnosics in more detail, especially for the different sections of the CFMT with increasing levels of difficulty. For the first section of the CFMT (identical images, for further details see (Duchaine and Nakayama, 2006b), the reliability of controls and prosopagnosics was similar (adjusted split-half coefficient[4]: prosopagnosics .63, controls .62, *one-tailed p* = .49). For the second section (novel images), prosopagnosics achieved significantly reduced reliability scores compared to controls (adjusted split-half coefficients: prosopagnosics .09, controls .74, *one-tailed p* = .042). For the third section (novel images with noise), prosopagnosics achieved marginally significantly reduced reliability scores compared to controls (adjusted split-half coefficients: prosopagnosics .32, controls .74, *one-tailed p* = .092). Based on these results, we give the following possible explanation. It is known that prosopagnosics use feature-based strategies to bypass their limited face recognition abilities in everyday life but also in test situations (Dalrymple et al., 2014; Mayer and Rossion, 2009; Grüter et al., 2011; Duchaine et al., 2003). The low reliability could be caused by this use of various strategies. Prosopagnosics might switch between strategies, combine several different strategies, or respond at random if they find that none of their strategies works. In our case, participants were not informed about the increasing difficulty levels in the CFMT. Prosopagnosics reported that they tried to remember single features of the faces. In the first section with identical images in learning and testing, this strategy proved beneficial. As soon as the target images became novel, prosopagnosics were unable to use their strategy any longer and had to guess sometimes. Another possible explanation could be that some internal processes for holistic face recognition do not work consistently for prosopagnosics. However, our tests results do not allow identifying the exact cause for this reduced reliability. Therefore, further testing is necessary, also to verify the robustness of this finding.

If indeed strategy usage, random answering, or inconsistent internal processes cause the reduced test reliability for prosopagnosics, this raises doubt whether the same perceptual processes and mechanisms are measured for controls and prosopagnosics. The same question also arises for the results of individual prosopagnosics. Because significant performance differences between controls and prosopagnosics were observed in at least one part of all face perception tests, we argue that these tests are suitable for a coarse comparison of face processing abilities between groups, even though for some tests there

---

[4] Split-half method was used for all sections, because Cronbach's alpha gave biased estimates due to ceiling performance of the controls in the first section

are apparently qualitative differences in reliability. However, for a more detailed analysis of performance levels, for example at an individual level, the tests might be too unreliable. In addition, the low reliabilities affect correlation analyses between tests. The correlation between test performances is restricted by the tests' reliabilities: the square root of the product of reliabilities of two tests gives an upper boundary to their correlation (Nunnally, 1970). Correlation analyses are often used to relate different face perception mechanisms, for example if holistic processing links to face identification performance (Konar et al., 2010; Richler et al., 2011; Degutis et al., 2013; Zhao et al., 2014). It is also used to examine if similar impairments exist in different individual cases of (Duchaine et al., 2007a; Kennerknecht et al., 2007; Duchaine et al., 2007b). Our finding therefore also has impact on the search for systematic patterns of impairment and possible common possible subgroups among prosopagnosics. As the low reliability for prosopagnosics adds noise to test results, this might complicate the identification of response patterns and subgroups in prosopagnosia, which is an actual focus of prosopagnosia research (Kress and Daum, 2003; Stollhoff et al., 2011; Esins et al., 2012)

6. General discussion

In the present study, we assessed and compared face perception of prosopagnosics to controls with several tests addressing different processes of face perception and object recognition. We investigated holistic processing of faces and object recognition, sensitivity to featural and configural facial information, gender recognition, benefit of motion information and the unconscious, automatic extraction of facial identity information. Each of the tests for face perception captured significant differences in performance between prosopagnosics and controls, but not the test for object recognition. We were able to replicate findings of previous studies, gain new insight about face processing impairments of prosopagnosics and resolve controversies raised by the conflicting results of previous studies. Furthermore, we investigated the internal reliability of these tests for both participant groups. We are the first to report that decreased reliability coefficients for prosopagnosic participants occur for tests of holistic processing of static faces.

Overall, our results confirmed that prosopagnosics, compared to controls, have a normal object recognition performance (4.2. Object recognition), but a reduced face recognition performance (4.1. CFMT, 4.3. Surprise recognition, and 4.7. Facial motion advantage). Furthermore, we could replicate the finding that prosopagnosics show limited holistic processing (4.4. Composite face task), and could refine this result by revealing that especially configural processing, and maybe featural processing to a lesser extent, are impaired (4.5. Featural and configural sensitivity). In addition, prosopagnosics lacked a facial motion advantage when faces were identical at learning and test, in contrast to control participants (4.7. Facial motion advantage). Prosopagnosics were impaired in gender recognition (4.6. Gender recognition), confirming previous studies that were based only on a few single cases of congenital prosopagnosia. Furthermore, we found that prosopagnosics showed automatic retrieval of identity information (4.3. Surprise recognition), indicating that their unconscious mechanisms for extracting identification-relevant information are still active whenever they view faces.

The reliability coefficients obtained for the controls and prosopagnosics reached good levels for most tests parts. However, prosopagnosics obtained only a fraction of the reliability coefficients of controls, exclusively for tests of holistic face processing using static stimuli. Especially for the CFMT prosopagnosics obtained a significantly lower reliability coefficient than controls. We suggest that this is caused by the fact that prosopagnosics use feature-based strategies to compensate for their face recognition impairment. This leads to their inconsistent answering behavior in tests of holistic processing of static faces.

This finding of reduced test reliabilities for prosopagnosics raises doubts whether the same perceptual processes are tested for controls and prosopagnosics in many face recognition tests. Therefore, in view of our reliability results, careful design of tests (to obtain higher reliability) and careful analyses of their results (to assess their reliability) are required. This is especially relevant when searching for possible subgroups in prosopagnosia defined by similar impairments in psychophysical tests.

Summary

With our extended battery of existing and newly created tests and our large sample size of prosopagnosic and control participants, we were able to shed further light into the face perception processes in congenital prosopagnosia. We replicated previous results of impaired face processing performance of prosopagnosics, and provided new results, deepening the understanding of prosopagnosia. Furthermore, we are the first to reveal that the response behavior of prosopagnosics in tests for holistic processing differs from controls, indicated by their noticeably reduced test reliability. Future work will need to examine the robustness and cause of this phenomenon. Additionally, better tests need to be designed, with higher reliabilities for prosopagnosics, providing more robust results to investigate prosopagnosia in more detail and obtain an accurate picture of the impairment.

References

Allison, T., Puce, A., and Mccarthy, G. (2000). Social perception from visual cues: role of the STS region. *Trends Cogn. Sci.* 4, 267–278. doi:10.1016/S1364-6613(00)01501-1.

Archer, J., Hay, D. C., and Young, A. W. (1992). Face processing in psychiatric conditions. *Br. J. Clin. Psychol.* 31, 45–61. doi:10.1111/j.2044-8260.1992.tb00967.x.

Ariel, R., and Sadeh, M. (1996). Congenital Visual Agnosia and Prosopagnosia in a Child: A Case Report. *Cortex* 32, 221–240. doi:10.1016/S0010-9452(96)80048-7.

Avidan, G., Tanzer, M., and Behrmann, M. (2011). Impaired holistic processing in congenital prosopagnosia. *Neuropsychologia* 49, 2541–2552. doi:10.1016/j.neuropsychologia.2011.05.002.

Avidan, G., Tanzer, M., Hadj-Bouziane, F., Liu, N., Ungerleider, L. G., and Behrmann, M. (2013). Selective Dissociation Between Core and Extended Regions of the Face Processing Network in Congenital Prosopagnosia. *Cereb. Cortex* 24, 1565–78. doi:10.1093/cercor/bht007.

Barton, J. J. S., Cherkasova, M. V, Hefter, R. L., Cox, T. A., O'Connor, M., and Manoach, D. S. (2004). Are patients with social developmental disorders prosopagnosic? Perceptual heterogeneity in the Asperger and socio-emotional processing disorders. *Brain* 127, 1706–16. doi:10.1093/brain/awh194.

Barton, J. J. S., Cherkasova, M. V, Press, D. Z., Intriligator, J. M., and O'Connor, M. (2003). Developmental prosopagnosia: A study of three patients. *Brain Cogn.* 51, 12–30. doi:10.1016/S0278-2626(02)00516-X.

Bate, S., Cook, S. J., Duchaine, B. C., Tree, J. J., Burns, E. J., and Hodgson, T. L. (2013). Intranasal Inhalation of Oxytocin Improves Face Processing in Developmental Prosopagnosia. *Cortex*, 1–9. doi:10.1016/j.cortex.2013.08.006.

Behrmann, M., and Avidan, G. (2005). Congenital prosopagnosia : face- blind from birth. *Trends Cogn. Sci.* 9, 180 – 187.

Behrmann, M., Avidan, G., Gao, F., and Black, S. (2007). Structural imaging reveals anatomical alterations in inferotemporal cortex in congenital prosopagnosia. *Cereb. Cortex* 17, 2354–63. doi:10.1093/cercor/bhl144.

Behrmann, M., Avidan, G., Marotta, J. J., and Kimchi, R. (2005). Detailed exploration of face-related processing in congenital prosopagnosia: 1. Behavioral findings. *J. Cogn. Neurosci.* 17, 1130–49. doi:10.1162/0898929054475154.

Bentin, S., Deouell, L. Y., and Soroker, N. (1999). Selective visual streaming in face recognition : evidence from developmental prosopagnosia. *Neuroreport* 10, 823–827.

Bodamer, J. (1947). Die Prosop-Agnosie. *Arch. für Psychiatr. und Nervenkrankheiten Ver. mit Zeitschrift für die Gesamte Neurol. und Psychiatr.* 179, 6–53. doi:10.1007/BF00352849.

Bonett, D. G. (2003). Sample Size Requirements for Comparing Two Alpha Coefficients. *Appl. Psychol. Meas.* 27, 72–74. doi:10.1177/0146621602239477.

Bowles, D. C., McKone, E., Dawel, A., Duchaine, B. C., Palermo, R., Schmalzl, L., Rivolta, D., Wilson, C. E., and Yovel, G. (2009). Diagnosing prosopagnosia: effects of ageing, sex, and participant-stimulus ethnic match on the Cambridge Face Memory Test and Cambridge Face Perception Test. *Cogn. Neuropsychol.* 26, 423–55. doi:10.1080/02643290903343149.

Brainard, D. H. (1997). The Psychophysics Toolbox. *Spat. Vis.* 10, 433–436. doi:10.1163/156856897X00357.

Brown, W. (1910). Some Experimental Results in the Correlation of Mental Abilities1. *Br. J. Psychol. 1904-1920* 3, 296–322.

Brunsdon, R., Coltheart, M., Nickels, L., and Joy, P. (2006). Developmental prosopagnosia: A case analysis and treatment study. *Cogn. Neuropsychol.* 23, 822–40. doi:10.1080/02643290500441841.

Chatterjee, G., and Nakayama, K. (2012). Normal facial age and gender perception in developmental prosopagnosia. *Cogn. Neuropsychol.* 29, 482–502. doi:10.1080/02643294.2012.756809.

Collishaw, S. M., and Hole, G. J. (2000). Featural and configurational processes in the recognition of faces of different familiarity. *Perception* 29, 893–909. doi:10.1068/p2949.

Crawford, J. R., Garthwaite, P. H., and Ryan, K. (2011). Comparing a single case to a control sample: testing for neuropsychological deficits and dissociations in the presence of covariates. *Cortex.* 47, 1166–78. doi:10.1016/j.cortex.2011.02.017.

Dalrymple, K. a., Fletcher, K., Corrow, S., Nair, R. Das, Barton, J. J. S., Yonas, A., and Duchaine, B. C. (2014). "A room full of strangers every day": The psychosocial impact of developmental prosopagnosia on children and their families. *J. Psychosom. Res.* doi:10.1016/j.jpsychores.2014.06.001.

Davidshofer, K. R., and Murphy, C. O. (2005). *Psychological testing: principles and applications.* 6th ed. Upper Saddle River, NJ: Pearson/Prentice Hal.

DeGutis, J. M., Chatterjee, G., Mercado, R. J., and Nakayama, K. (2012). Face gender recognition in developmental prosopagnosia: Evidence for holistic processing and use of configural information. *Vis. cogn.* 20, 1242–1253. doi:10.1080/13506285.2012.744788.

DeGutis, J. M., Cohan, S., and Nakayama, K. (2014). Holistic face training enhances face processing in developmental prosopagnosia. *Brain* 137, 1781–98. doi:10.1093/brain/awu062.

Degutis, J. M., Wilmer, J. B., Mercado, R. J., and Cohan, S. (2013). Using regression to measure holistic face processing reveals a strong link with face recognition ability. *Cognition* 126, 87–100. doi:10.1016/j.cognition.2012.09.004.

Dennett, H. W., McKone, E., Tavashmi, R., Hall, A., Pidcock, M., Edwards, M., and Duchaine, B. C. (2011). The Cambridge Car Memory Test: A task matched in format to the Cambridge Face Memory Test, with norms, reliability, sex differences, dissociations from face memory, and expertise effects. *Behav. Res. Methods.* doi:10.3758/s13428-011-0160-2.

Dobel, C., Bölte, J., Aicher, M., and Schweinberger, S. R. (2007). Prosopagnosia without apparent cause: Overview and diagnosis of six cases. *Cortex* 2, 718–733.

Dobs, K., Bülthoff, I., and Schultz, J. (2015). Identity information in facial motion varies with the type of facial movement. *Manuscr. Prep.*

Duchaine, B. C., Germine, L. T., and Nakayama, K. (2007a). Family resemblance: ten family members with prosopagnosia and within-class object agnosia. *Cogn. Neuropsychol.* 24, 419–30. doi:10.1080/02643290701380491.

Duchaine, B. C., and Nakayama, K. (2006a). Developmental prosopagnosia: a window to content-specific face processing. *Curr. Opin. Neurobiol.* 16, 166–73. doi:10.1016/j.conb.2006.03.003.

Duchaine, B. C., and Nakayama, K. (2006b). The Cambridge Face Memory Test: results for neurologically intact individuals and an investigation of its validity using inverted face stimuli and prosopagnosic participants. *Neuropsychologia* 44, 576–85. doi:10.1016/j.neuropsychologia.2005.07.001.

Duchaine, B. C., Parker, H., and Nakayama, K. (2003). Normal recognition of emotion in a prosopagnosic. *Perception* 32, 827–838. doi:10.1068/p5067.

Duchaine, B. C., Yovel, G., Butterworth, E., and Nakayama, K. (2006). Prosopagnosia as an impairment to face-specific mechanisms: Elimination of the alternative hypotheses in a developmental case. *Cogn. Neuropsychol.* 23, 714–747. doi:10.1080/02643290500441296.

Duchaine, B. C., Yovel, G., and Nakayama, K. (2007b). No global processing deficit in the Navon task in 14 developmental prosopagnosics. *Soc. Cogn. Affect. Neurosci.* 2, 104–13. doi:10.1093/scan/nsm003.

Eddy, J. K., and Glass, A. L. (1981). Reading and listening to high and low imagery sentences. *J. Verbal Learning Verbal Behav.* 20, 333–345. doi:10.1016/S0022-5371(81)90483-7.

Esins, J., Bülthoff, I., Kennerknecht, I., and Schultz, J. (2012). Can a test battery reveal subgroups in congenital prosopagnosia? *Percept. 41 ECVP Abstr. Suppl.* 41, 113–113.

Esins, J., Bülthoff, I., and Schultz, J. (2014a). Motion does not improve face recognition accuracy in congenital prosopagnosia. *J. Vis. 2014 VSS Abstr. Suppl.* 14, 1436–1436. doi:10.1167/14.10.1436.

Esins, J., Bülthoff, I., and Schultz, J. (2011). The role of featural and configural information for perceived similarity between faces. *J. Vis. 2011 VSS Abstr. Suppl.* 11, 673–673. doi:10.1167/11.11.673.

Esins, J., Schultz, J., Bülthoff, I., and Kennerknecht, I. (2013). Galactose uncovers face recognition and mental images in congenital prosopagnosia: The first case report. *Nutr. Neurosci.* 0, 1–2. doi:10.1179/1476830513Y.0000000091.

Esins, J., Schultz, J., Wallraven, C., and BÃ¼lthoff, I. (2014b). Do congenital prosopagnosia and the other-race effect affect the same face recognition mechanisms? *Front. Hum. Neurosci.* 8, 1–14. doi:10.3389/fnhum.2014.00759.

Farah, M. J., Wilson, K. D., Drain, M., and Tanaka, J. W. (1998). What is "special" about face perception? *Psychol. Rev.* 105, 482–98. Available at: http://www.ncbi.nlm.nih.gov/pubmed/9697428.

Farah, M. J., Wilson, K. D., Maxwell Drain, H., and Tanaka, J. R. (1995). The inverted face inversion effect in prosopagnosia: Evidence for mandatory, face-specific perceptual mechanisms. *Vision Res.* 35, 2089–2093. doi:10.1016/0042-6989(94)00273-O.

Fisher, R. A. (1921). On the "Probable Error" of a Coefficient of Correlation Deduced from a Small Sample. *Metron* 1, 3–32.

Freire, A., Lee, K., and Symons, L. a (2000). The face-inversion effect as a deficit in the encoding of configural information: Direct evidence. *Perception* 29, 159–170. doi:10.1068/p3012.

Garrido, L., Duchaine, B. C., and Nakayama, K. (2008). Face detection in normal and prosopagnosic individuals. *J. Neuropsychol.* 2, 119–140. doi:10.1348/174866407X246843.

Garrido, L., Furl, N., Draganski, B., Weiskopf, N., Stevens, J., Tan, G. C.-Y., Driver, J., Dolan, R. J., and Duchaine, B. C. (2009). Voxel-based morphometry reveals reduced grey matter volume in the temporal cortex of developmental prosopagnosics. *Brain* 132, 3443–55. doi:10.1093/brain/awp271.

De Gelder, B., Bachoud-Lévi, A.-C., and Degos, J.-D. (1998). Inversion superiority in visual agnosia may be common to a variety of orientation polarised objects besides faces. *Vision Res.* 38, 2855–2861. doi:10.1016/S0042-6989(97)00458-6.

Gomez, J. L., Pestilli, F., Witthoft, N., Golarai, G., Liberman, A., Poltoratski, S., Yoon, J., and Grill-Spector, K. (2015). Functionally Defined White Matter Reveals Segregated Pathways in Human Ventral Temporal Cortex Associated with Category-Specific Processing. *Neuron* 85, 216–227. doi:10.1016/j.neuron.2014.12.027.

Le Grand, R., Cooper, P. A., Mondloch, C. J., Lewis, T. L., Sagiv, N., De Gelder, B., and Maurer, D. (2006). What aspects of face processing are impaired in developmental prosopagnosia? *Brain Cogn.* 61, 139–58. doi:10.1016/j.bandc.2005.11.005.

Gruber, T., Dobel, C., Jungho, M., and Junghöfer, M. (2011). The Role of Gamma-Band Activity in the Representation of Faces: Reduced Activity in the Fusiform Face Area in Congenital Prosopagnosia. *PLoS One* 6, e19550. doi:10.1371/journal.pone.0019550.

Grüter, M. (2004). Genetik der kongenitalen Prosopagnosie [Genetics of congenital prosopagnosia]. *Unpubl. Dr. Thesis, Medizinische Fak. der Westfälischen Wilhelms-Universität Münster, Münster.*

Grüter, M., Grüter, T., Bell, V., Horst, J., Laskowski, W., Sperling, K., Halligan, P. W., Ellis, H. D., and Kennerknecht, I. (2007). Hereditary Prosopagnosia: the first case series. *Cortex* 43, 734–749. Available at: http://thomasgrueter.de/Grueter_et_al_2007cortex.pdf [Accessed January 17, 2012].

Grüter, T., Grüter, M., and Carbon, C.-C. (2011). Congenital prosopagnosia. Diagnosis and mental imagery: commentary on "Tree JJ, and Wilkie J. Face and object imagery in congenital prosopagnosia: a case series.". *Cortex.* 47, 511–3. doi:10.1016/j.cortex.2010.08.005.

Grüter, T., Grüter, M., and Carbon, C.-C. (2008). Neural and genetic foundations of face recognition and prosopagnosia. *J. Neuropsychol.* 2, 79–97. doi:10.1348/174866407X231001.

De Haan, E. H. F., and Campbell, R. (1991). A fifteen year follow-up of a case of developmental prosopagnosia. *Cortex* 27, 489–509. doi:10.1016/s0010-9452(13)80001-9.

Haxby, J. V., Hoffman, E. A., and Gobbini, M. I. (2000). The distributed human neural system for face perception. *Trends Cogn. Sci.* 4, 223–233. Available at: http://www.ncbi.nlm.nih.gov/pubmed/10827445.

Hayward, W. G., Rhodes, G., and Schwaninger, A. (2008). An own-race advantage for components as well as configurations in face recognition. *Cognition* 106, 1017–27. doi:10.1016/j.cognition.2007.04.002.

Herzmann, G., Danthiir, V., Schacht, A., Sommer, W., and Wilhelm, O. (2008). Toward a comprehensive test battery for face cognition: Assessment of the tasks. *Behav. Res. Methods* 40, 840–857. doi:10.3758/BRM.40.3.840.

Hoffman, E. A., and Haxby, J. V. (2000). Distinct representations of eye gaze and identity in the distributed human neural system for face perception. *Nat. Neurosci.* 3, 80–4. doi:10.1038/71152.

IBM Corp. Released 2011. IBM SPSS Statistics for Windows, Version 20.0. Armonk, NY: IBM Corp.

Ishai, A., Schmidt, C. F., and Boesiger, P. (2005). Face perception is mediated by a distributed cortical network. *Brain Res. Bull.* 67, 87–93. doi:10.1016/j.brainresbull.2005.05.027.

Jones, R. D., and Tranel, D. (2001). Severe Developmental Prosopagnosia in a Child With Superior Intellect. *J. Clin. Exp. Neuropsychol.* 23, 265–273(9). doi:http://dx.doi.org/10.1076/jcen.23.3.265.1183.

Kanwisher, N., and Yovel, G. (2006). The fusiform face area: a cortical region specialized for the perception of faces. *Philos. Trans. R. Soc. Lond. B. Biol. Sci.* 361, 2109–28. doi:10.1098/rstb.2006.1934.

Kaulard, K., Cunningham, D. W., Bülthoff, H. H., and Wallraven, C. (2012). The MPI facial expression database--a validated database of emotional and conversational facial expressions. *PLoS One* 7, e32321. doi:10.1371/journal.pone.0032321.

Kennerknecht, I. (2011, 9 February). Prosopagnosie oder das Problem, Gesichter wieder zu erkennen. Available at: http://www.prosopagnosia.de/ [Accessed January 8, 2015].

Kennerknecht, I., Grüter, T., Welling, B., and Wentzek, S. (2006). First Report of Prevalence of Non-Syndromic Hereditary Prosopagnosia (HPA). *Am. J. Med. Genet.*, 1617 – 1622. doi:10.1002/ajmg.a.

Kennerknecht, I., Kischka, C., Stemper, C., Elze, T., and Stollhoff, R. (2011). "Heritability of face recognition," in *Face Analysis, Modeling and Recognition Systems*, ed. T. Barbu (InTech), 163–188. Available at: http://gendocs.ru/docs/18/17804/conv_1/file1.pdf#page=175 [Accessed July 14, 2014].

Kennerknecht, I., Plümpe, N., Edwards, S., and Raman, R. (2007). Hereditary prosopagnosia (HPA): the first report outside the Caucasian population. *J. Hum. Genet.* 52, 230–6. doi:10.1007/s10038-006-0101-6.

Kennerknecht, I., Welling, B., and Pluempe, N. (2008). Congenital prosopagnosia – a common hereditary cognitive dysfunction in humans. *Front. Biosci.* 13, 3150–3158.

Kimchi, R., Behrmann, M., Avidan, G., and Amishav, R. (2012). Perceptual separability of featural and configural information in congenital prosopagnosia. *Cogn. Neuropsychol.* 29, 447–63. doi:10.1080/02643294.2012.752723.

Kleiner, M., Brainard, D., and Pelli, D. (2007). What's new in Psychtoolbox-3? in *Perception 36 ECVP Abstract Supplement.*

Konar, Y., Bennett, P. J., and Sekuler, A. B. (2010). Holistic processing is not correlated with face-identification accuracy. *Psychol. Sci.* 21, 38–43. doi:10.1177/0956797609356508.

Kress, T., and Daum, I. (2003). Developmental prosopagnosia: a review. *Behav. Neurol.* 14, 109–21. doi:10.1155/2003/520476.

Lance, C. E., Butts, M. M., and Michels, L. C. (2006). The Sources of Four Commonly Reported Cutoff Criteria. *Organ. Res. Methods* 9, 202–220.

Lange, J., de Lussanet, M., Kuhlmann, S., Zimmermann, A., Lappe, M., Zwitserlood, P., and Dobel, C. (2009). Impairments of biological motion perception in congenital prosopagnosia. *PLoS One* 4, e7414. doi:10.1371/journal.pone.0007414.

Lee, H.-J., Macbeth, A. H., Pagani, J., and Young, W. S. 3rd (2009). Oxytocin: the Great Facilitator of Life. *Prog. Neurobiol.* 88, 127–151. doi:10.1016/j.pneurobio.2009.04.001.Oxytocin.

Lobmaier, J. S., Bölte, J., Mast, F. W., and Dobel, C. (2010). Configural and featural processing in humans with congenital prosopagnosia. *Adv. Cogn. Psychol.* 6, 23–34. doi:10.2478/v10053-008-0074-4.

Longmore, C. A., and Tree, J. J. (2013). Motion as a cue to face recognition: Evidence from congenital prosopagnosia. *Neuropsychologia* 51, 1–12. doi:10.1016/j.neuropsychologia.2013.01.022.

Malpass, R. S., and Kravitz, J. (1969). Recognition for faces of own and other race. *J. Pers. Soc. Psychol.* 13, 330–4. Available at: http://www.ncbi.nlm.nih.gov/pubmed/5359231.

Markovska-Simoska, S. M., and Pop-Jordanova, N. (2010). Face and emotion recognition by ADHD and normal adults. *Acta Neuropsychol.* 8, 99–122.

Marks, D. F. (1973). Visual imagery differences in the recall of pictures. *Br. J. Psychol.* 64, 17–24. Available at: http://www.ncbi.nlm.nih.gov/pubmed/4742442.

Maurer, D., Le Grand, R., and Mondloch, C. J. (2002). The many faces of configural processing. *Trends Cogn. Sci.* 6, 255–260. doi:10.1016/S1364-6613(02)01903-4.

Maurer, D., O'Craven, K. M., Le Grand, R., Mondloch, C. J., Springer, M. V, Lewis, T. L., and Grady, C. L. (2007). Neural correlates of processing facial identity based on features versus their spacing. *Neuropsychologia* 45, 1438–51. doi:10.1016/j.neuropsychologia.2006.11.016.

Mayer, E., and Rossion, B. (2009). "Prosopagnosia," in *The Behavioral and Cognitive Neurology of Stroke*, eds. O. Godefroy and J. Bogousslavsky (Cambridge: Cambridge University Press), 316–335. Available at: https://www.yumpu.com/en/document/view/12830258/prosopagnosia.

McKone, E., Davies, A. A., Darke, H., Crookes, K., Wickramariyaratne, T., Zappia, S., Fiorentini, C., Favelle, S., Broughton, M., and Fernando, D. (2013). Importance of the inverted control in measuring holistic face processing with the composite effect and part-whole effect. *Front. Psychol.* 4, 33. doi:10.3389/fpsyg.2013.00033.

McKone, E., Hall, A., Pidcock, M., Palermo, R., Wilkinson, R. B., Rivolta, D., Yovel, G., Davis, J. M., and O'Connor, K. B. (2011). Face ethnicity and measurement reliability affect face recognition performance in developmental prosopagnosia: evidence from the Cambridge Face Memory Test-Australian. *Cogn. Neuropsychol.* 28, 109–46. doi:10.1080/02643294.2011.616880.

McKone, E., Stokes, S., Liu, J., Cohan, S., Fiorentini, C., Pidcock, M., Yovel, G., Broughton, M., and Pelleg, M. (2012). A robust method of measuring other-race and other-ethnicity effects: the Cambridge Face Memory Test format. *PLoS One* 7, 1–6. doi:10.1371/journal.pone.0047956.

Meissner, C. A., and Brigham, J. C. (2001). Thirty years of investigating the own-race bias in memory for faces: A meta-analytic review. *Psychol. Public Policy, Law* 7, 3–35. doi:10.1037//1076-8971.7.1.3.

Michel, C., Caldara, R., and Rossion, B. (2006a). Same-race faces are perceived more holistically than other-race faces. *Vis. cogn.* 14, 55–73. doi:10.1080/13506280500158761.

Michel, C., Rossion, B., Han, J., Chung, C.-S., and Caldara, R. (2006b). Holistic processing is finely tuned for faces of one's own race. *Psychol. Sci.* 17, 608–15. doi:10.1111/j.1467-9280.2006.01752.x.

Mosetter, K. (2008). "Chronischer Streß auf der Ebene der Molekularbiologie und Neurobiochemie," in *Psychodynamische Psycho-und Traumatherapie* (VS Verlag für Sozialwissenschaften), 77–98.

Nieminen-von Wendt, T., Paavonen, J. E., Ylisaukko-Oja, T., Sarenius, S., Källman, T., Järvelä, I., and von Wendt, L. (2005). Subjective face recognition difficulties, aberrant sensibility, sleeping disturbances and aberrant eating habits in families with Asperger syndrome. *BMC Psychiatry* 5, 20. doi:10.1186/1471-244X-5-20.

Nunnally, J. C. (1970). *Introduction to psycological measurement.* New York: McGraw-Hill.

Nunnally, J. C. (1978). *Psychometric theory.* 2nd ed. New York: McGraw-Hill.

O'Toole, A. J., Harms, J., Snow, S. L., Hurst, D. R., Pappas, M. R., Ayyad, J. H., and Abdi, H. (2005). A video database of moving faces and people. *IEEE Trans. Pattern Anal. Mach. Intell.* 27, 812–6. doi:10.1109/TPAMI.2005.90.

Palermo, R., Willis, M. L., Rivolta, D., McKone, E., Wilson, C. E., and Calder, A. J. (2011). Impaired holistic coding of facial expression and facial identity in congenital prosopagnosia. *Neuropsychologia* 49, 1226–35. doi:10.1016/j.neuropsychologia.2011.02.021.

Pitcher, D., Walsh, V., and Duchaine, B. C. (2011). The role of the occipital face area in the cortical face perception network. *Exp. brain Res.* 209, 481–93. doi:10.1007/s00221-011-2579-1.

Pyles, J. a, Verstynen, T. D., Schneider, W., and Tarr, M. J. (2013). Explicating the face perception network with white matter connectivity. *PLoS One* 8, e61611. doi:10.1371/journal.pone.0061611.

Rhodes, G., Brake, S., Taylor, K., and Tan, S. (1989). Expertise and configural coding in face recognition. *Br. J. Psychol.* 80, 313–331. doi:10.1111/j.2044-8295.1989.tb02323.x.

Richler, J. J., Cheung, O. S., and Gauthier, I. (2011). Holistic processing predicts face recognition. *Psychol. Sci.* 22, 464–471. doi:10.1177/0956797611401753.Holistic.

Rivolta, D., Palermo, R., Schmalzl, L., and Coltheart, M. (2011). Covert face recognition in congenital prosopagnosia: A group study. *Cortex* 48, 1–9. doi:10.1016/j.cortex.2011.01.005.

Rivolta, D., Palermo, R., Schmalzl, L., and Williams, M. a (2012). Investigating the features of the m170 in congenital prosopagnosia. *Front. Hum. Neurosci.* 6, 45. doi:10.3389/fnhum.2012.00045.

Rodrigues, A., Bolognani, S. A. P. S., Brucki, S. S. M. D., and Bueno, O. F. A. O. (2008). Developmental prosopagnosia and adaptative compensatory strategies. *Dement. Neuropsychol.* 2, 353–355. Available at: http://www.demneuropsy.com.br/imageBank/PDF/dnv02n04a20.pdf [Accessed September 29, 2014].

Russell, R., Duchaine, B. C., and Nakayama, K. (2009). Super-recognizers: people with extraordinary face recognition ability. *Psychon. Bull. Rev.* 16, 252–7. doi:10.3758/PBR.16.2.252.

Salkovic-Petrisic, M., Osmanovic-Barilar, J., Knezovic, A., Hoyer, S., Mosetter, K., and Reutter, W. (2014). Long-term oral galactose treatment prevents cognitive deficits in male Wistar rats treated intracerebroventricularly with streptozotocin. *Neuropharmacology* 77, 68–80. doi:10.1016/j.neuropharm.2013.09.002.

Savaskan, E., Ehrhardt, R., Schulz, A., Walter, M., and Schächinger, H. (2008). Post-learning intranasal oxytocin modulates human memory for facial identity. *Psychoneuroendocrinology* 33, 368–74. doi:10.1016/j.psyneuen.2007.12.004.

Schmalzl, L., Palermo, R., and Coltheart, M. (2008a). Cognitive heterogeneity in genetically based prosopagnosia: A family study. *J. Neuropsychol.* 2, 99–117. doi:10.1348/174866407X256554.

Schmalzl, L., Palermo, R., Green, M., Brunsdon, R., and Coltheart, M. (2008b). Training of familiar face recognition and visual scan paths for faces in a child with congenital prosopagnosia. *Cogn. Neuropsychol.* 25, 704–29. doi:10.1080/02643290802299350.

Schwarzer, G., Huber, S., and Dümmler, T. (2005). Gaze behavior in analytical and holistic face processing. *Mem. Cognit.* 33, 344–354. doi:10.3758/BF03195322.

Schweich, M., and Bruyer, R. (1993). Heterogeneity in the cognitive manifestations of prosopagnosia: The study of a group of single cases. *Cogn. Neuropsychol.* 10, 529–547. doi:10.1080/02643299308253472.

Shah, P., Gaule, A., Gaigg, S. B., Bird, G., and Cook, R. (2015). Probing short-term face memory in developmental prosopagnosia. *Cortex* 64, 115–122. doi:10.1016/j.cortex.2014.10.006.

Spearman, C. (1910). Correlation calculated from faulty data. *Br. J. Psychol. 1904-1920* 3, 270–295.

Stollhoff, R. (2010). Modeling Prosopagnosia: Computational Theory and Experimental Investigations of a Deficit in Face Recognition (Doctoral dissertation, Phd thesis). Available at: http://www.fmi.unileipzig.de/promotion/abstract.stollhoff.pdf.

Stollhoff, R., Jost, J., Elze, T., and Kennerknecht, I. (2011). Deficits in long-term recognition memory reveal dissociated subtypes in congenital prosopagnosia. *PLoS One* 6, e15702. doi:10.1371.

Susilo, T., McKone, E., Dennett, H. W., Darke, H., Palermo, R., Hall, A., Pidcock, M., Dawel, A., Jeffery, L., Wilson, C. E., et al. (2010). Face recognition impairments despite normal holistic processing and face space coding: Evidence from a case of developmental prosopagnosia. *Cogn. Neuropsychol.* 27, 636–664. doi:10.1080/02643294.2011.613372.

Tanaka, J. W., and Farah, M. J. (1993). Parts and wholes in face recognition. *Q. J. Exp. Psychol. A.* 46, 225–45. Available at: http://www.ncbi.nlm.nih.gov/pubmed/8316637.

The MathWorks Inc. MATLAB and Statistics Toolbox Release 2011b. Natick, Massachusetts, United States.

Thomas, C., Avidan, G., Humphreys, K., Jung, K.-J., Gao, F., and Behrmann, M. (2009). Reduced structural connectivity in ventral visual cortex in congenital prosopagnosia. *Nat. Neurosci.* 12, 29–31. doi:10.1038/nn.2224.

Tree, J. J., and Wilkie, J. (2010). Face and object imagery in congenital prosopagnosia: a case series. *Cortex* 46, 1189–98. doi:10.1016/j.cortex.2010.03.005.

Wilmer, J. B., Germine, L., Chabris, C. F., Chatterjee, G., Williams, M., Loken, E., Nakayama, K., and Duchaine, B. C. (2010). Human face recognition ability is specific and highly heritable. *Proc. Natl. Acad. Sci. U. S. A.* 107, 5238–41. doi:10.1073/pnas.0913053107.

Xiao, N. G., Quinn, P. C., Ge, L., and Lee, K. (2013). Elastic facial movement influences part-based but not holistic processing. *J. Exp. Psychol. Hum. Percept. Perform.* 39, 1457–67. doi:10.1037/a0031631.

Young, A. W., Hellawell, D., and Hay, D. C. (1987). Configurational information in face perception. *Perception* 16, 747–759. Available at: http://www.perceptionweb.com/perception/fulltext/p16/p160747.pdf [Accessed February 20, 2014].

Yovel, G., and Duchaine, B. C. (2006). Specialized face perception mechanisms extract both part and spacing information: evidence from developmental prosopagnosia. *J. Cogn. Neurosci.* 18, 580–93. doi:10.1162/jocn.2006.18.4.580.

Zhao, M., Hayward, W. G., and Bülthoff, I. (2014). Holistic processing, contact, and the other-race effect in face recognition. *Vision Res.* 105, 61–69. doi:10.1016/j.visres.2014.09.006.

III. Do congenital prosopagnosia and the other-race effect affect the same face recognition mechanisms?

1. Abstract

Congenital prosopagnosia (CP), an innate impairment in recognizing faces, as well as the other-race effect (ORE), a disadvantage in recognizing faces of foreign races, both affect face recognition abilities. Are the same face processing mechanisms affected in both situations? To investigate this question, we tested three groups of 21 participants: German congenital prosopagnosics, South Korean participants and German controls on three different tasks involving faces and objects. First we tested all participants on the Cambridge Face Memory Test in which they had to recognize Caucasian target faces in a 3-alternative-forced-choice task. German controls performed better than Koreans who performed better than prosopagnosics. In the second experiment, participants rated the similarity of Caucasian faces that differed parametrically in either features or second-order relations (configuration). Prosopagnosics were less sensitive to configuration changes than both other groups. In addition, while all groups were more sensitive to changes in features than in configuration, this difference was smaller in Koreans. In the third experiment, participants had to learn exemplars of artificial objects, natural objects, and faces and recognize them among distractors of the same category. Here prosopagnosics performed worse than participants in the other two groups only when they were tested on face stimuli. In sum, Koreans and prosopagnosic participants differed from German controls in different ways in all tests. This suggests that German congenital prosopagnosics perceive Caucasian faces differently than do Korean participants. Importantly, our results suggest that different processing impairments underlie the ORE and CP.

2.   Introduction

Recognizing faces is arguably the most important way to identify other humans and bears great social importance. Even though faces are a visually homogeneous object class, most humans are experts in face identification: within milliseconds we can identify a familiar face in poor lighting, after 15 years of aging, 20 pounds of weight loss, or with a different hairdo - and this is true for the several hundred acquaintances we have on average.

One explanation for this achievement is that we use "holistic processing" for faces: we integrate the different components of a face [e.g., the form and color of the features (eyes, nose, and mouth) and their configuration (i.e., spatial distances between the features)] into a whole and do not process single pieces of information individually (Maurer et al., 2002). If the retrieval of this information is disturbed, holistic processing and thus face recognition are impaired (Collishaw and Hole, 2000). Especially configural processing is considered to be one of the most important aspects of holistic processing: disturbing this process alone already strongly affects holistic processing of faces (Maurer et al., 2002).

Most humans are undoubtedly experts at every-day face recognition but this expertise can be disturbed in various ways. Two well-known phenomena in which people show impaired face recognition abilities are congenital prosopagnosia (CP) and the other-race effect (ORE).

CP is an innate impairment in face processing. People with CP often encounter social difficulties, like being considered arrogant or ignorant because they fail to recognize and greet acquaintances. Therefore, some of them tend to keep a socially withdrawn life. Presumably 2.5 % of the population is affected (Kennerknecht et al., 2008). In contrast to the acquired form of prosopagnosia, which is caused by acquired brain damage, CP is inborn and there are no evident brain lesions. Also several studies found normal functional brain response to faces in fMRI studies (e.g., Avidan et al., 2005; Avidan and Behrmann, 2009) and EEG studies (e.g., Towler et al., 2012) but subtle differences in connectivity between face processing brain regions for congenital prosopagnosics compared with controls (Avidan et al., 2008). In a single case study of CP, this reduced connectivity could be enhanced by training on spatial integration of mouth and eye regions of faces. The training also had positive effects on face recognition performance but vanished after a few months (DeGutis et al., 2007).

The ORE describes the fact that we recognize faces of our own (familiar) race faster and more accurately than faces of an unfamiliar ethnicity (Meissner and Brigham, 2001). This effect (also called "cross-race bias," "own-race advantage," or "other- race deficit") is a common and known phenomenon. Several models exist to explain the underlying mechanisms causing the ORE. The most common explanation is the higher level of expertise for same-race faces compared with other-race faces (Meissner and Brigham, 2001). This perceptual expertise hypothesis states that the frequent encounter and the training in individuating own-race faces leads to a greater experience in encoding the dimensions most useful to individuate faces of that race. Nevertheless, competing models exists, like the social categorization hypothesis, which states that mere social out-group categorization is sufficient to elicit a drop in face recognition performance (Bernstein et al., 2007). Another hypothesis is the categorization-individuation model which combines perceptual experience, social categorization and motivated individuation (discrimination among individuals within a racial group which requires attending to face-identity characteristics rather than to category-diagnostic characteristics), all three of which co-act and generate the ORE (Hugenberg et al., 2010). The underlying mechanisms are not clear yet, but it has been shown that the ORE can be overcome by training, but only for the trained faces (McKone et al., 2007).

As nearly everyone has experienced the ORE, it is sometimes cited as an example by congenital prosopagnosics when they try to describe to non-prosopagnosics what they experience in everyday life. Both phenomena are characterized by the difficulty in telling people apart or recognizing previously encountered people based on their faces. But also, in both cases, there is evidence for parallels in disturbances of face processing as reviewed in the following.

Some studies used the inversion effect or the composite face effect to test face processing abilities of their participants. The inversion effect describes the effect that face recognition performance is reduced if the faces are presented upside down. The strength of this effect is significantly larger for faces than for other objects for which we are not experts. The composite face effect describes the illusion of a new identity when combining the top half of the face of one person with the bottom half face of another person. The two halves cannot be processed individually and create the face of a new, third person. The illusion disappears when the two halves are misaligned. Both effects, the face inversion and the composite face effect, are considered to be hallmarks for holistic face processing. Both disrupt the configural information leaving the featural information intact. This again is an indication of the importance of configural processing for holistic processing (Maurer et al., 2002). A study testing congenital prosopagnosic participants found no face inversion effect or composite-face effect, neither in accuracy nor in reaction times, indicating their impairment in holistic processing of faces (Avidan et al., 2011). Regarding the face inversion effect for other-race faces, two experiments testing European and Asian participants found a larger effect for same-race faces than for other-race faces in both groups of participants (Rhodes et al., 1989). When testing the composite face task with Asian and European participants, similarly, Michel and colleagues found a significantly larger composite face effect for same-race faces compared with other-race faces (Michel et al., 2006).

In a study conducted by Lobmaier and colleagues, congenital prosopagnosics were tested with scrambled faces (configural information destroyed) and blurred faces (featural information destroyed) in a delayed matching task. Prosopagnosic participants showed significantly worse performance than controls in both conditions (Lobmaier et al., 2010). Chinese and Caucasian- Australian participants tested in an old-new recognition task on blurred and scrambled Asian and Caucasian faces also showed a significantly worse performance for other-race faces than for own-race faces in both conditions (Hayward et al., 2008).

In another study, congenital prosopagnosics participants were tested on a same-different task with the so-called "Jane" set of stimuli (Le Grand et al., 2006). These stimuli faces differ either in features, configuration, or contour. Only a minority of the prosopagnosic participants performed significantly worse than controls on the faces differing in configuration or features, but most prosopagnosics performed significantly worse on faces differing in their contour. A study with Asian participants using the same "Jane" stimuli and a similarly created Asian female face set also showed only marginal effects (Mondloch et al., 2010): Chinese participants were significantly slower on other-race compared with same-race faces (analysis collapsed over all three types (features, configuration, contour), with the longest mean reaction times for the faces differing in contour) but showed no significant differences in performance for any modification (features, configuration, contour). Even though this lack of differences between groups for the "Jane" stimuli was challenged by (Yovel and Duchaine, 2006) (this will be discussed in our general discussion), we note that similar results for other-race observers and prosopagnosic observers were obtained in both studies.

There are several different causes that can reduce face recognition ability (aging, illnesses, drug consumption, etc.). However, the two face recognition disturbances under study here, CP and the ORE, seem to impair face recognition abilities in a similar way, namely by disrupting featural and configural face processing (depending on the used stimuli and task, as reviewed above) causing a lack or reduction of face expertise. Also, in both cases face recognition performance can be increased to a certain extent through training. These similarities could be a hint that the same face processing mechanisms are impaired.

To verify the hypothesis of a common underlying disturbance, it is necessary to compare in detail whether the same kind of impairments appear when looking specifically and directly at featural and configural processing. On one hand, if differences in face recognition performance appear, we can exclude a common underlying disturbance. On the other hand, if similar impairments are found, the hypothesis that the same mechanisms are disturbed is not proven, but possible. In any case, a direct comparison between CP and the ORE is a great chance to get further insights into the yet unknown mechanisms underlying face processing and face recognition.

To conduct this direct comparison we recruited three age- and gender-matched participant groups with a comparatively large sample size of 21 participants per group: German congenial prosopagnosic participants, Korean participants, and German controls. All participant groups performed the same three tests. (1) the Cambridge Face Memory Test (CFMT, Duchaine and Nakayama, 2006), an objective measure of the face recognition abilities of Caucasian faces, (2) a parametric test of the sensitivity to configural and featural information in faces; sensitivity to these two types of facial information has been shown to be reduced in congenital prosopagnosics and other-race observers in previous studies, and (3) a recognition task of faces and familiar and unfamiliar objects to test the influence of expertise on recognition performance.

As all face stimuli used in our tests were derived from Caucasian faces, we expected the Korean group to exhibit evidence of the ORE that could be compared with the performance of the prosopagnosics while the German control group would serve as a baseline. Our predictions for each test were the following: (1) For the CFMT, Koreans and prosopagnosics would have a lower score compared with German controls, due to the disadvantage in recognizing other-race faces for the Koreans and the innate face recognition impairment for the prosopagnosics. This test is a general measure of the severity of face recognition impairments and does not detect if differences in the nature of the impairments exist. (2) We expected to find a decreased sensitivity to configural and featural information for prosopagnosics and Koreans. This prediction was based on reported deficits in processing both kinds of information in prosopagnosic as well as other-race observers (Hayward et al., 2008; Lobmaier et al., 2010 respectively). If prosopagnosics and Koreans would show differences in the extraction of featural and configural information, we could exclude that common mechanisms are impaired. (3) In the object and face recognition test we expected an impaired recognition performance of the face stimuli for Koreans and prosopagnosics, again due to the disadvantage in recognizing other-race faces for the Koreans and the innate face recognition impairment for the prosopagnosics. We expected to find no differences across all participant groups in recognizing the non-expertise object stimuli. Despite a study describing that 54 congenital prosopagnosics self- reported impaired object recognition during interviews (Grüter et al., 2008), most studies explicitly testing object recognition found nearly-normal to normal object recognition abilities for prosopagnosic participants. When impairments were found, they were less pronounced than face recognition impairments (see Kress and Daum, 2003; Le Grand et al., 2006 for reviews).

3.   Materials and methods

3.1.   Participants

We tested three groups of participants: German congenital prosopagnosic participants (from now on referred to as "prosopagnosics"), South Korean participants ("Koreans"), and German control participants ("Germans") with 21 participants per group. The ratio of female to male participants as well as the age of participants in each group was matched as closely as possible. Note that it was hard to recruit older male Korean participants, presumably for cultural reasons; therefore we had to resort to younger male participants in that group to have matching numbers of participants in all groups.

So far, no universally-accepted standard diagnostic tool for CP exists: while the CFMT is widely used to characterize prosopagnosic participants (e.g., Rivolta et al., 2011; Kimchi et al., 2012), other diagnostic means exist. The prosopagnosics of our study were identified by a questionnaire and interview (Stollhoff et al., 2011). Due to time constraints the Koreans and Germans did not participate in the diagnostic interview but reported to have no problems in recognizing faces of their friends and family members. To provide an objective measure of face processing abilities and to maintain comparability with other studies, we tested all participants on the CFMT and report their scores and z-scores, based on the results of the German controls, in Table 1.

German congenital prosopagnosic participants

The prosopagnosics were diagnosed by the Institute of Human Genetics, Universitäts-klinikum Münster, based on a screening questionnaire and an diagnostic semi-structured interview (Stollhoff et al., 2011). All prosopagnosics were tested at the Max Planck Institute for Biological Cybernetics in Tübingen, Germany and compensated with 8 Euro per hour plus travel expenses.

Korean participants

The Korean participants were compensated with 30,000 Won (approximately 20 Euro) for the whole experiment. All participants of this group were tested at Korea University in Seoul, South Korea. The Koreans did not perform a diagnostic interview but were asked if they had noticeable problems recognizing faces of friends and family members. None of the participants reported face recognition impairments.

German control participants

The German control participants were compensated with 8 Euro per hour. All participants of this group were tested at the Max Planck Institute for Biological Cybernetics in Tübingen, Germany. The Germans did not perform a diagnostic interview but were asked if they had noticeable problems recognizing faces of friends and family members. None of the participants reported face recognition impairments.

Table 1: Overview of the participants in the three different groups.

	Prosopagnosic				Korean				German			
	Sex	Age	CFMT		Sex	Age	CFMT		Sex	Age	CFMT	
			score	z-score			score	z-score			score	z-score
1	f	21	38	-3.57	f	22	53	-1.05	f	23	65	0.96
2	f	22	44	-2.57	f	23	53	-1.05	f	24	69	1.63
3	f	24	37	-3.74	m	24	47	-2.06	f	24	64	0.79
4	f	27	47	-2.06	m	24	57	-0.38	f	25	57	-0.38
5	f	27	42	-2.90	m	26	51	-1.39	f	29	61	0.29
6	f	28	36	-3.91	f	28	57	-0.38	f	31	53	-1.05
7	m	33	45	-2.40	m	30	50	-1.56	m	33	59	-0.05
8	m	34	33	-4.41	m	37	53	-1.05	f	36	55	-0.72
9	f	36	38	-3.57	f	39	58	-0.22	m	36	58	-0.22
10	m	36	45	-2.40	m	41	55	-0.72	m	37	50	-1.56
11	m	37	34	-4.24	f	41	55	-0.72	f	37	64	0.79
12	f	41	34	-4.24	f	42	53	-1.05	m	39	62	0.46
13	f	46	44	-2.57	f	42	63	0.62	m	39	52	-1.22
14	f	46	39	-3.40	f	45	64	0.79	m	44	71	1.97
15	m	47	43	-2.73	f	46	44	-2.57	f	44	52	-1.22
16	m	52	40	-3.24	f	50	47	-2.06	f	46	59	-0.05
17	f	53	36	-3.91	f	51	63	0.62	f	47	54	-0.89
18	f	54	46	-2.23	f	55	54	-0.89	f	49	58	-0.22
19	m	57	37	-3.74	f	55	38	-3.57	m	54	68	1.46
20	m	59	38	-3.57	f	57	50	-1.56	f	58	54	-0.89
21	f	64	38	-3.57	f	58	50	-1.56	m	60	60	0.12
Mean scores			39.17	-3.28			53.10	-1.04			59.29	0.00
♂	8				6				8			
Mean age		40.2				39.8				38.8		

Depicted are their sex (f, female; m, male), age in years, and their scores in the CFMT as well as the according z-scores, based on the results of the German controls.

All participants provided informed consent. All participants have normal or corrected-to-normal visual acuity.

3.2. Analysis

Many studies found faster reaction times for Asian compared with Caucasian participants regardless of the task (Rushton and Jensen, 2005). We made similar observations in our study and hence we do not compare reaction times between our Asian and Caucasian participants, as any comparison would not give interpretable results. Nevertheless, we compared reaction times for prosopagnosics and Germans for the object recognition task, as participants in both groups share the same ethnicity.

All analyses were conducted with Matlab2011b (Natick, MA) and IBM SPSS Statistics Version 20 (Armonk, NY). The dependent variables analyzed in each test are described in the respective sections.

We report effect sizes as partial eta square (η^2_p). For One-Way ANOVAs partial eta square and eta square (η^2) are the same. For our Two-Way ANOVAs partial eta square differs from eta square, therefore we give both values.

3.3. Apparatus

All participants were tested individually. For prosopagnosics and Germans the experiments were run on a desktop PC with 24" screen, Koreans performed the tests on a MacBook Pro with a 17" screen. The CFMT is Java-script based; Matlab and Psychtoolbox were used to run the other experiments. Participants were seated at a viewing distance of approximately 60cm from the screen.

3.4. Procedure

The procedure was approved by the local IRB. All participants completed three tests: (1) the CFMT, (2) a rating task of the similarity of faces differing in features or configuration, (3) an object recognition task. All tests were conducted in the same order to obtain comparable results for each participant. Participants could take self-paced breaks between experiments.

4. Test battery

4.1. Cambridge face memory test

Motivation

The CFMT was created and provided by Bradley Duchaine and Ken Nakayama (Duchaine and Nakayama, 2006). This test assesses recognition abilities using unfamiliar faces in a 3- alternative-forced-choice task. It has been widely used in recent years in studies of CP and of the ORE. Therefore, we used it here as an objective measure of face recognition abilities.

Stimuli

As this test has been described in detail in the original study, only a short description is given here. Pictures of the faces of young male Caucasians shown under three different viewpoints and under different lighting and noise conditions were used in recognition tests of increasing difficulty. For a complete description of the test see the original study (Duchaine and Nakayama, 2006).

Task

First the participants were familiarized with six target faces which they then had to recognize among distractors in a 3-alternative- forced-choice task with tests of increasing difficulty. No feedback was given. The test can be run in an upright and inverted condition. We only used the upright condition.

Results

The percent correct recognition of participants was calculated and the mean and standard error of the three participants groups are depicted in Figure 1.

Germans (mean percent correct = 82.3 %, SD = 8.3) performed significantly better than Koreans (mean = 73.7 %, SD = 8.8), who performed significantly better than prosopagnosics (mean = 55.2 %, SD = 5.9) [One-Way ANOVA: $F(2, 62) = 67.34$, $p < .001$, $\eta^2_p = .69$, with Tukey HSD post-hoc tests: all comparisons $p \leq .002$].

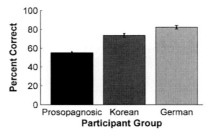

Figure 1: Performance of the 3 participant groups in the CFMT. Data are displayed as mean percentage correct responses. Error bars: SEM.

As predicted, the Koreans and prosopagnosics performed significantly worse than the Germans. Furthermore, the prosopagnosics performed significantly worse than the Koreans. The significant difference in performance for the Germans and Koreans shows an own-race advantage for the Germans. We assume that reduced performance of the Koreans is due to the ORE; however, as we did not perform the reverse test with Asian faces, we cannot completely exclude an alternative cause for this difference between participant groups. We suggest that this is very unlikely, because the CFMT and its Chinese version (comprising Chinese faces depicted in a similar way and format as the faces in the CFMT; only published after our data acquisition) were already success- fully used to measure the ORE in a complete cross-over design in Caucasian and Asian participants (McKone et al., 2012).

From our finding that Koreans show a significantly better recognition performance than prosopagnosics we cannot exclude that the same mechanisms for processing Caucasian faces are affected in these groups. But we can infer that CP has a stronger impact on face recognition abilities compared with the ORE.

4.2.   Similarity rating of faces differing in features and configuration

Motivation
This test was conducted to measure in what way and to what extent the retrieval of featural and configural information is disturbed in other-race observers and

prosopagnosics. Based on this pattern we want to infer if we can exclude that the same mechanisms for processing Caucasian faces are affected in CP and the ORE. As discussed in the introduction, previous studies found disturbances in holistic processing (e.g., Avidan et al., 2011 for CP; Rhodes et al., 1989; Michel et al., 2006 for the ORE), and disruptions of configural and featural processing (e.g., Lobmaier et al., 2010 for CP; Hayward et al., 2008 for the ORE). However, other studies using different tasks and stimuli found only minor or no impairments in configural and featural processing (e.g., Le Grand et al., 2006 for CP; Mondloch et al., 2010 for the ORE). The pattern of findings obtained so far was too inconsistent and not detailed enough to draw conclusions regarding our research question. To resolve this controversy and to obtain usable data, we assessed the fine-grained sensitivity to featural and configural facial information and compared the effects of CP and the ORE.

Stimulus creation
We generated eight natural-looking face sets with gradual small- step changes in features and configuration to determine the grade of sensitivity to featural and configural facial information, without resorting to unnatural modifications (like blurring or scrambling). The faces in each of our stimulus sets differ only in internal features and their configuration. Skin texture and outer face shape were held constant to allow testing purely for sensitivity to internal features and configuration. The face stimuli contain no extra-facial cues (no hair, makeup, clothing, or jewelry).

The stimuli were created using faces from our in-house 3D face database (Troje and Bülthoff, 1996). The faces are 3D laser scans of the faces of real persons. A morphable model allows to isolate and exchange the four main face regions between any faces of the database (Vetter and Blanz, 1999). Those four regions are: both eyes (including eyebrows), the nose, the mouth, and the outer face shape (Figure 2). For these regions, the texture (i.e., "skin") and / or the shape can be morphed as well as exchanged between all faces. Additionally the regions can be shifted within each face (e.g., moving the eyes up or apart of each other).

We chose pairs of faces from the database such that the faces in each pair differed largely from each other in both configuration and features. Previous studies that have used faces differing in either features or configuration have shown that participants are more sensitive to featural than to configural changes (Freire et al., 2000; Goffaux et al., 2005;

Maurer et al., 2007; Rotshtein et al., 2007). For this reason we further increased the configural differences of the face pairs by shifting the features slightly (e.g., we moved the eyes closer together in the face which had more closely spaced eyes, and moved the eyes further apart in the other face of the pair). This was done for best conditions to measure configural sensitivity, as this is one main focus of our study, while remaining within natural limits. That the faces are still perceived as natural was tested in a pilot study described further below.

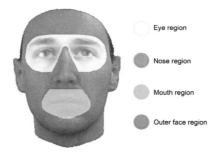

Figure 2: Illustration of the editable regions of the 3D faces of our in-house face database (Troje and Bülthoff, 1996).

The outer face shape and skin texture of the modified faces were averaged within each pair and applied to both modified faces to create two faces A and B (Figure 3B). A and B exhibit different features and inner configuration but identical averaged outer face shape and skin texture. Based on the faces A and B we then generated two more faces by creating a face X with features of face A and the configuration of face B (i.e., the features of face A were moved to the feature locations of face B) and vice versa for face Y (see scheme in Figure 3A; see actual face stimuli in Figure 3B). By morphing between these four faces in 25 % increments we generated a whole set of faces parametrically differing from each other in features (Figure 3C, horizontal axes) or configuration (Figure 3C, vertical axes). We created eight different sets in the same way as the one depicted in Figure 3C,one for each of eight pairs of original faces of our database (note: each original face was used only in one set).

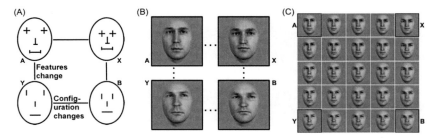

Figure 3: (A) Schematic four faces which either differ in features (horizontal) or configuration (vertical). (B) The same design is applied to real faces of our face database. (C) Morphing between the four faces in (B) gives a set. Morphing steps between each row and column are equally spaced with 25 %.

To ensure that the faces we created appeared just as natural as the original faces, we ran a pilot study in which participants rated the naturalness of the modified and original faces without any knowledge about the facial modifications. The modified faces we used for our study showed no significant difference in perceived naturalness compared with the original scanned faces of real people (Esins et al., 2011).

Further, to verify that featural and configural modifications introduced similar amounts of changes in the pictures, we calculated the mean pixelwise image differences between the stimuli with the greatest configural and featural parametrical differences per set. We took the two end point faces of the vertical bar (see Figure 4) and calculated their Euclidean distance for each pixel and did the same for the two end point faces of the horizontal bar. Then we calculated the average pixel distance for the two comparisons[1]. With this method we obtained mean Euclidean pixel distances for configural and featural changes, for each of the eight created sets. A Wilcoxon signed rank test run on all eight mean distances for the featural changes vs. the eight configural change distances was not significant (p = .31), supporting the idea that featural and configural face modifications introduced similar amounts of computational change in the pictures.

---

[1] Only pixels which actually differed between both images were taken into consideration. Thus, the gray background and the common outer face shape were omitted for the averaging process. This avoids an artificial reduction of the mean pixel distances.

Configuration changes

Features change

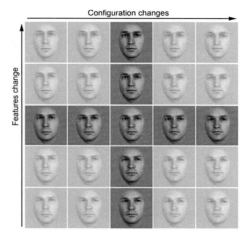

Figure 4: One of the eight sets of face stimuli used in the similarity rating experiment. Only faces of the central horizontal and vertical bars were used for the experiments. The endpoint faces were used to calculate mean pixelwise image differences between the stimuli.

Task

Participants had to rate the pair-wise similarity of faces originating from the same set. Due to time limitations we used only nine test faces per set: the ones located on the central horizontal bar (differing in features) and the central vertical bar (differing in configuration) of each set (see Figure 4). Each face was compared with the eight other faces on the central bars of the same set and with itself. Trials in which faces differed in both, features and configuration, were considered filler trials to avoid participants realizing the nature of the stimuli and were omitted from the analysis. Therefore, in sum, for each of the eight sets, we analyzed 29 pair-wise similarity ratings: nine identical face comparisons (100 % parametrical similarity), eight face comparisons with 75 % parametrical similarity (two faces next to each other in the set), six face comparisons with 50 % parametrical similarity, four face comparisons with 25 % parametrical similarity, and two face comparisons with 0 % parametrical similarity (comparison of the extreme faces of the same bar). So in total there were 232 comparisons during this experiment. The order of comparisons was randomized within and across sets for each participant.

Participants had to rate the perceived similarity on a Likert scale from 1 (little similarity) to 7 (high similarity/identical) and were told to use the whole range of ratings over the whole experiment. The participants saw the first face for 2000 ms, then a pixelated face mask for 800 ms, and then the second face for another 2000 ms. Subsequently, the Likert scale appeared on the screen: here participants marked their rating by moving a slider via the arrow keys on the keyboard (Figure 5). The start position of the slider was randomized. There was no time restriction for entering the answer, however, participants were told to rate the similarity without too long considerations. After every 20 comparisons there was a self-paced pause.

The face and mask stimuli had a size of approximately 5.7° horizontal and 8.6° vertical visual angle. To prevent pixel matching, the faces were presented at different random positions on the screen within a viewing angle of about 7.6° horizontally and 10.5° vertically.

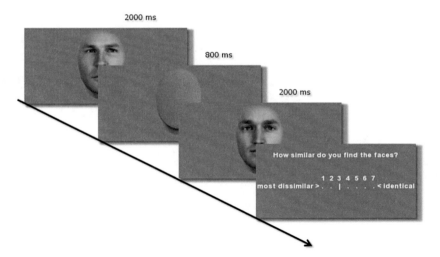

Figure 5: Example of one trial of the similarity rating task. Both faces in a trial always belong to the same set.

Analysis

For every participant we calculated the mean similarity ratings across all eight sets at each of the five levels of parametric similarity (100, 75, 50, 25, 0 %). Example data of one German participant is given in Figure 6. The black triangles show the average rating of face pairs of all sets differing in features, sorted by the different parametrical similarities. The gray squares show the same for configural changes. As expected, Germans gave similarity ratings close to 7 (high similarity) for very similar faces.

A linear regression ($y = \beta x + \varepsilon$) was fitted to these mean similarity ratings (dotted black and gray lines in Figure 6). The steepness of the slopes (β) was then used as a measure of sensitivity: steeper slopes indicate more strongly perceived configural or featural changes. For every participant we calculated one regression slope for their featural and one for their configural ratings. The mean and the standard error of the sensitivity β per participant group are illustrated in Figure 7A.

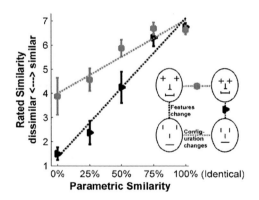

Figure 6: Exemplar results of one German participant of the similarity ratings. For each of the five similarity levels, the average ratings across all face comparisons of all sets were calculated. The sensitivity ratings for changes in features (black triangles) and configuration (gray squares) are shown separately. The error bars depict standard error. A linear regression ($y = \beta x + \varepsilon$) was fitted to both curves individually (dotted black and dotted gray, respectively). The slopes (β) serve as measure of the sensitivity to features and configuration.

To compare performance data, we took a closer look at the pattern of sensitivity to features and configuration: For each individual participant, we subtracted their configural sensitivity from their featural sensitivity. We refer to this difference as 'featural advantage'. The illustration in Figure 7B shows the mean of the calculated differences, i.e., the mean of the featural advantage for each group.

Results

A 2×3 ANOVA on the regression slopes β as a measure of sensitivity showed that the main effect of change type (configural, featural) was significant [$F(1, 60) = 233.7$, $p < .001$, $\eta^2 = .46$, $\eta^2_p = .796$]. All participants showed a greater sensitivity to changes in features than to changes in configurations. The main effect of participant group (prosopagnosics, Koreans, Germans) was also significant [$F(2, 60) = 6.46$, $p = .003$, $\eta^2 = .07$, $\eta^2_p = .18$]. The interaction between change type and participant group was significant, too [$F(2, 0) = 5.48$, $p = .007$, $\eta^2 = .02$, $\eta^2_p = .15$].

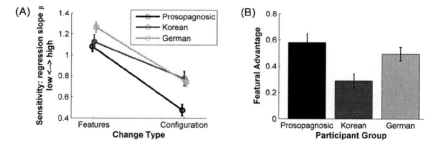

Figure 7: Results of the similarity rating experiment. (A) Mean values of slopes (β) for the "feature" and "configuration" regression lines for each group Error bars: SEM. (B) "featural advantage": mean difference between configural and featural regression slopes (β) calculated for each participant. Error bars: SEM.

Analysis of simple effects for both change types (configural, featural) was carried out: The group differences of sensitivity to features approaches significance [One-Way ANOVA $F(2, 62) = 3.12$, $p = .0515$, $\eta^2_p = .09$], which was mainly driven by the difference between prosopagnosic and Germans (Tukey HSD post-hoc test, $p = .051$, both other differences $p > .17$). For configural changes there were significant group differences in sensitivity [One-Way ANOVA $F(2, 62) = 9.11$, $p < .001$, $\eta^2_p = .23$] with prosopagnosics performing significantly differently from Koreans and Germans (Tukey HSD post-hoc test, $p = .001$ and $p = .003$, respectively. Tukey HSD post-hoc test for Koreans vs. Germans $p = .91$).

For analysis of the featural advantage (Figure 7B) we conducted a One-Way ANOVA to further examine the significant interaction of the main effects (participant group vs. change type). The ANOVA showed significant differences between the three groups [$F(2, 62) = 5.48$, $p = .007$, $\eta^2_p = .15$], which are the same values as for the interaction in the 2×3ANOVA,as expected. The Tukey HSD post-hoc tests revealed significant differences in the featural advantage between Koreans and prosopagnosics ($p = .005$), a difference approaching significance for the Koreans vs. the Germans ($p = .091$) and no difference for prosopagnosics vs. Germans ($p = .51$).

Discussion

There is a clear difference in sensitivity to features and configuration of our stimuli faces between Koreans and prosopagnosics: while both groups show about the same sensitivity to featural changes, we found that prosopagnosics have a significantly reduced sensitivity to configuration compared with Koreans (and Germans). Also the featural advantage was significantly smaller for Koreans than for the prosopagnosics. These differences in absolute sensitivity to configural and featural changes, and also the differences in featural advantage, suggest that Korean and prosopagnosic participants do not perceive our Caucasian face stimuli in the same way. Because CP and the ORE show parallels in disrupting featural and configural face processing, we hypothesized that the same mechanisms are disturbed in both cases. This would result in a similarly reduced sensitivity to features and configuration for participants affected by CP or the ORE. But as Korean and prosopagnosic participants show a different pattern of disturbance of their sensitivity, we can reject this hypothesis and conclude that different underlying mechanisms are affected. Our similarity rating task also allowed to obtain a more detailed picture of the sensitivities to featural and configural information in CP and the ORE. For the prosopagnosics compared with the Germans, the difference between both groups approached significance for sensitivity to features and reached significance for sensitivity to configuration (Figure 7A). Our results show a marginally significant difference for prosopagnosics and Germans in featural sensitivity ($p = .051$). These results bridge the gap between two studies reporting conflicting results using the so-called "Jane" stimuli (Le Grand et al., 2006) and "Alfred" stimuli (Yovel and Kanwisher, 2004; Yovel and Duchaine, 2006), which, like our stimuli, also differ in features and configuration (and contour for the "Jane" stimuli). Only a minority of the prosopagnosic participants performed significantly worse than controls on the "Jane" stimuli differing in features and configuration (Le Grand et al., 2006). Based on the data by Le Grand and colleagues given in Table 4 of that study, comparing prosopagnosics and controls, one can estimate that there was a significant performance difference for the configural but not for the featural modifications. Yovel and colleagues also used the "Jane" stimuli with prosopagnosics and controls and confirmed the significant performance difference between groups for configural modifications and non-significant difference for featural modifications (Yovel and Duchaine, 2006). However, they challenged the "Jane" stimuli for including obvious brightness differences (due to makeup) for the featural modifications. For their own "Alfred" stimuli they found significantly reduced sensitivity to featural and configural modifications for prosopagnosic participants (Yovel and Duchaine, 2006; Duchaine et al., 2007). In turn, their "Alfred" stimuli were challenged for configural modifications going beyond natural limits (as discussed in Maurer et al., 2007). Our newly created stimulus set contains no extra-facial cues (no hair, makeup, glasses, or beard) and exhibits configural changes which have been tested to be within natural limits. With these well controlled stimuli our results suggest that for prosopagnosic participants, the retrieval of the configural information of a face is indeed impaired compared with the Germans. For the sensitivity to features, our results lie between the non-significant results obtained with the "Jane" stimuli and the significant results obtained with "Alfred" faces. Therefore, we conclude that the retrieval of featural information might be impaired for prosopagnosics, although to a lesser degree than the retrieval of configural information.

We found no significant difference in sensitivity to featural or configural information between the Korean and German groups. Our results are in concordance with a previous study, also using the "Jane" stimuli, that found no differences between Caucasian and Asian participants (Mondloch et al., 2010). In contrast, other studies found an own-race advantage for both configuration and feature changes (Rhodes et al., 2006; Hayward et al., 2008). However, we note that the stimuli used in those latter studies involved different kinds of changes than those used in our present study (features and configuration were changed by blurring and scrambling (Hayward et al., 2008) or features were changed through changes in color (Rhodes et al., 2006), which opens the possibility that the ORE impacts differently on the perception of these different kinds of stimulus modifications. Nevertheless, as our stimuli contain more natural and ecological modifications of faces, we believe that our results better reflect participants' face perception. Even though we found no significant differences in sensitivity to featural or configural information between Germans and Koreans, we found that the featural advantage shows a trend to be larger for the Germans compared with the Koreans. Although this difference only approaches significance, we present two explanations for this pattern. The first explanation is that due to the ORE, the sensitivity pattern is altered for our Korean participants. The ORE could reflect Koreans' lower expertise with other-race facial features whereas their configural processing stays unaffected when viewing other-race faces. The second explanation is that the effect is due to cultural differences. Studies have shown that Western Caucasian and Eastern Asian participants focus at different areas of faces and have dissimilar patterns of fixation when looking at faces (Blais et al., 2008). It might be that German and Korean participants employ different strategies when comparing faces in our task, which could have caused the effects we found. In accordance with this hypothesis, a study using Navon figures reported that Eastern Asian participants focus more on global configuration compared with Western Caucasian participants (McKone et al., 2010). By analogy, a greater focus on configurations in faces could explain the reduced featural advantage we observed in the Korean group.

Furthermore, our results show that all groups, regardless of their race and face recognition abilities, were more sensitive to differences in the featural than in the configural dimension of our stimulus set (Figure 7A). The presence of a featural advantage is in accordance with findings of previous studies using faces modified within natural limits in their configuration and features, where participants showed a higher sensitivity for featural changes as well (Freire et al., 2000; Goffaux et al., 2005; Maurer et al., 2007; Rotshtein et al., 2007). Even though for the "Alfred" stimuli similar sensitivities to featural and configural modifications were found by Yovel and Kanwisher (2004),their result should be regarded with caution in view of the unnatural configural modifications of their face stimuli (as discussed in Maurer et al., 2007). In contrast, we took care that our face stimuli were always natural looking and pixelwise analyses of our stimuli, as described earlier, have revealed no differences in induced image changes in the featural and configural dimensions. In other words, our stimuli exhibit the same pixelwise variation for the featural and configural changes. The fact that the observers nevertheless show a featural advantage suggests that humans are more sensitive to featural information, and/or perceive these changes to be more profound than changes in configuration. Another possible explanation is that it is more difficult to compare faces differing in configuration than to compare faces differing in features. Additionally, differences between two naturally-occurring faces are more likely to be featural than configural. Therefore, the human face discrimination system might have developed to be better at detecting featural than featural differences between faces.

4.3. Object recognition

Motivation
In this test we measured the influence of expertise on recognition performance. To this end, we compared recognition performance for objects for which one group has expertise (Caucasian faces) to recognition performance for objects for which no group has expertise (seashells and blue objects).

Stimulus creation
Three categories of stimuli were used: computer renditions of natural objects (seashells), artificial novel objects (blue objects, dissimilar to any known shapes) and faces. See Figure 8 for examples of these three categories of objects. All objects and faces where full 3D models, allowing to train and test participants on different viewpoints (see below). For each category we created four targets and twelve distractors.

Sixteen synthetic seashells were taken from a previously created stimulus set (Gaißert et al., 2010). The shells were created using a mathematical model (Fowler et al., 1992) implemented in the software ShellyLib (www.shelly.de). Attention was paid to sample stimuli spread evenly over the parametrically defined stimulus set space (see Gaißert et al., 2010 for details).

The blue objects were created with 3D Studio Max by Christoph D. Dahl (unpublished work) and were novel to all participants. Differences between these objects are less obvious for a human observer, making recognition more difficult.

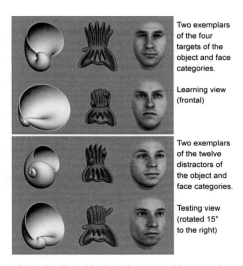

Two exemplars of the four targets of the object and face categories.

Learning view (frontal)

Two exemplars of the twelve distractors of the object and face categories.

Testing view (rotated 15° to the right)

Figure 8: Exemplars of the stimuli used in the object recognition experiment.

For the face stimuli, 16 male Caucasian faces were selected from the MPI 3D face database (Troje and Bülthoff, 1996). The 16 faces where chosen to have as little salient distinctive features as possible (all were clean shaven, had the same gaze direction; showed no blemishes or moles, etc.).

None of the stimuli had been seen before by our participants. We created two sets of images for each stimulus category: frontal views for the learning phase, and stimuli rotated by 15 degrees to the right around the vertical axis (yaw) for the testing phase. The change between learning and testing was designed to prevent pixel matching of the stimuli.

All stimuli were shown at a viewing angle of approximately 9.5° horizontally and vertically.

Task

There was one block of trials per stimulus category, with the same procedure in all three blocks, as follows: During the learning phase, participants had to memorize four target exemplars depicted in frontal view. First, all four targets were shown together on the screen, then each of the four targets was shown one after the other, and finally all target exemplars were presented together again. Participants could control when to switch to the next screen via a button press. They were aware that if they switched to the next view they could not return to the previous one. No time restriction was applied. During testing, participants saw the images depicting the targets and distractors of the same category under a new orientation and performed an old-new-decision task by pressing buttons on a standard computer keyboard (old = left hand button press; new = right hand button press).Stimuli were presented for a duration of 2000 ms or until key press, whichever came first. The next image appeared as soon as an answer was entered.

Targets and distractors were presented in pseudo-randomized order: The testing was divided into three runs. Four targets and four distractors per category were shown in each run. While the targets were the same in each run, four new distractors were presented, such that all four targets were seen three times and each of the 12 distractors was seen only once. The order of the stimulus blocks (shells, faces then blue objects) was fixed to induce similar effects of tiredness in all participants. Participants took short self-paced breaks between blocks.

We kept the number of targets and distractors low, as performing tests with faces can be demotivating for prosopagnosics. We used the same number of stimuli in all stimulus categories to ensure comparability. The high similarity between the non- face objects was designed to avoid ceiling performance despite the low number of stimuli and to mimic the homogeneity of the face stimuli.

Analysis

The results were analyzed based on the dependent measure d'. The term d' refers to signal-detection theory measures (Macmillan and Creelman, 2005) and is an index of subjects' ability to dis- criminate between signal (target stimuli) and noise (distractors). The maximum possible d' value in this experiment is 3.46 (this depends on the number of trials). A d' of zero indicates chance discrimination performance, higher values indicate increasing ability to tell targets and distractors apart.

Results

For a summary analysis of the general influence of object category (faces, shells, blue objects) and participant group (prosopagnosics, Koreans, Germans) we ran a 3×3 ANOVA on the d' values. The main effect of participant group was not significant [$F(2, 60) = 1.22, p = .303, \eta^2 = .009, \eta^2_p = .04$] but the main effect of object category was [$F(2, 60) = 145.54, p < .001, \eta^2 = .52, \eta^2_p = .71$], as well as the interaction between participant group and object category [$F(4, 120) = 7.14, p < .001, \eta^2 = .05, \eta^2_p = .19$]. Figure 9 depicts the performance of all groups graphically. The Germans and the Koreans were better at recognizing faces than shells and worst for recognizing the blue objects. This order differs for the prosopagnosics who were best at recognizing shells, faces and blue objects in that order.

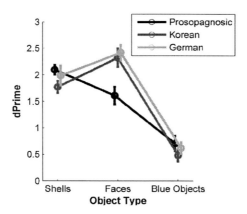

Figure 9: Performance of the three participant groups in the object recognition task. Data are shown as mean d' values. Error bars: SEM.

A One-Way ANOVA on the d' values for each object cate- gory across participant groups revealed significant differences for the face stimuli: $F(2, 62) = 8.14, p = .001, \eta^2_p = .21$. A post-hoc analysis showed that prosopagnosics' performance was significantly different from the other two groups (Games Howel test, $p \leq .01$ for prosopagnosics vs. Koreans and prosopagnosics vs. Germans). The other One-Way ANOVAs and post-hoc tests on the level of shells and blue objects, respectively, were not significant (all $ps > .2$). We also compared reaction times of Germans and prosopagnosics for the non-face object categories (shells, blue objects)with the Wilcoxon Rank sum test. We found no significant differences ($p = .13$ for shells, $p = .31$ for blue objects).

Discussion

As expected, no significant differences between groups were found for shells and blue objects. This can be explained by the fact that all participants, equally, were non-experts for these objects. Performance differed only for faces. We found that prosopagnosics, as non-experts for faces, performed less well on face recognition than the other two groups. Interestingly, the Koreans, also non-experts for our Caucasian stimuli, did not exhibit a lower recognition performance than Germans. An obvious reason for the absence of the ORE is the small amount of targets to be memorized for this test. It is thus likely that the task was too easy for all non-prosopagnosic participants. For the prosopagnosics, our results show that the task is difficult even with this small amount of target faces. This confirms the results we observed in the CFMT, namely that CP has a stronger impact on face recognition abilities compared with the ORE.

We compared recognition performance for faces not only with one type of objects but with easy and difficult object categories, which reduces the risk of ceiling or flooring effects. Germans and Koreans recognized the non-face objects less easily than the faces, probably because, even for Koreans, their expertise for faces is better than their expertise for the visually similar non- face objects. For prosopagnosics the accuracy performance for faces lay between their performance for easy and difficult object categories. This indicates that the stimuli were not too easy to recognize.

Our findings confirm previous results indicating that, although some prosopagnosics might show object recognition deficits, those impairments are less severe than their face recognition deficits (Kress and Daum, 2003; Le Grand et al., 2006). But a further aspect of object recognition expertise worth exploring is reaction times. Behrmann and colleagues found that object recognition deficit of their five prosopagnosic participants does not show in accuracy performance, but in reaction time (Behrmann et al., 2005); and in a study by Duchaine and Nakayama (2005), many prosopagnosic participants exhibited longer reaction times rather than lower recognition accuracy compared with control participants: four of their seven prosopagnosic participants had a reaction time slower by more than 2 SD compared with the mean reaction time of their controls in most tasks. We did not find slower reaction times for non-face object recognition for prosopagnosics compared with Germans. These results thus exclude a general recognition deficit in our prosopagnosics.

5. Correlations between tests

Given that we ran several face processing experiments with different tasks testing for different aspects of recognition, we also examined the degree of correlation between test performances. For this we calculated Pearson's correlations between task performances across participants of all groups (Table 2).

Table 2: correlations between tests across all participants

N=63	Sensitivity to		Object and face recognition			
	Feat	Conf	Faces	Shells	Blue Objects	
CFMT	**0.30**	**0.49**	**0.48**	-0.10	-0.05	Correlation coefficient
	(0.016)	(0.000)	(0.000)	(0.45)	(0.71)	(p-value)
Feat		**0.48**	**0.33**	0.01	0.05	
		(0.000)	(0.009)	(0.94)	(0.72)	
Conf			0.22	**-0.30**	-0.16	
			(0.078)	(0.016)	(0.21)	
Faces				0.21	0.16	
				(0.10)	(0.21)	
Shells					0.18	
					(0.17)	

Depicted are the correlation coefficient, and in parentheses the p-value of the coefficient. Negative correlations are marked in red, significant correlations are written in bold letters. (CFMT, final score; Feat, sensitivity to featural changes in a face; Conf, sensitivity to configural changes in a face; Shells, Faces, Blue objects: d' values for shells, faces and the blue objects in the object recognition task.)

Performance on all four face-related tasks [CFMT, sensitivity to features (Feat) and configuration (Conf), object recognition task with face stimuli (Faces)] were positively and significantly correlated or approached significance. The effect sizes of these correlations ($.22 < r < .49$) were medium and hence the proportions of shared variance ($.05 < r^2 < .24$) were rather small. Thus, we assume that although different aspects of face perception are investigated by the tests (i.e., recognition performance, memory, and sensitivity to features and configuration) these aspects are nevertheless to some degree dependent from each other.

Surprisingly there was another significant, but negative correlation (with a rather small effect size): participants with a high sensitivity to configuration of a face tended to have bad performance in the shell recognition task. The small proportion of shared variance of $r^2 = .09$ led us to refrain from any speculations.

6.   General discussion

The combination of tasks used in this study tested various aspects of face and object recognition, which allowed us to compare directly the influence of CP and the ORE. Our hypothesis, based on previous findings, was that in CP and the ORE the same underlying mechanisms might be affected. While we could dis- prove this hypothesis (this is discussed in detail below), we were able to confirm results of previous studies and importantly we gain new insights concerning the similarities between these two impairments of face recognition.

First, we were able to replicate the findings that congenital prosopagnosics exhibit face recognition deficits but no object recognition deficits (Le Grand et al., 2006). Second, we were able to replicate the ORE with our Koreans in the CFMT. Interestingly our results differ somewhat from the results by McKone et al. (2012) who only found a trend toward a different performance between their Asian and Caucasian participants on the original CFMT. A possible explanation for this discrepancy is that their Asian participants may have had more experience with Caucasian faces because they were overseas students living in Australia at the time of testing. Our Asian participants were tested in Korea and thus were likely to have less experience with Caucasian faces. Third, our experiment testing sensitivity toward featural and configural changes within a face resolves discrepancies between studies testing sensitivity toward featural and configural facial information for prosopagnosics (Le Grand et al., 2006; Yovel and Duchaine, 2006). Our results, in the context of previous studies, show that, compared with German controls, prosopagnosics exhibit an impaired sensitivity toward configural information and possibly and only to a lesser extent, toward featural information of a face.

Importantly, besides those confirmations of previous findings, we report the new finding that sensitivities to features and configuration of a face differ between Korean and prosopagnosic participants. For both groups, the observed sensitivity to the featural changes in a face was about the same. The Koreans, however, were better than prosopagnosics (and as good as Germans) at detecting fine changes in configural information in a face. When comparing CP with the ORE, we asked if they derive from a disturbance in the same underlying mechanisms. Our results indicate that this is not the case: especially the difference in absolute sensitivity to configural and featural changes for prosopagnosic and other-race observers is a strong indicator that CP and the ORE impair face recognition differently. As we used the same face stimuli to test all participant groups, our results indicate that lacking expertise for a certain face group does not impact configural processing of those faces (Korean group),while CP does (prosopagnosic group). Even though we cannot explain what exactly causes this difference, these results clearly show that there are different mechanisms underlying both impairments. Therefore, we are not "prosopagnosic for other-race faces" (see also Wang et al., 2009).

Our second main finding is that face recognition performance is more strongly affected by CP than by the ORE. Our prosopagnosics performed significantly worse than the Koreans in all face recognition tasks. A possible explanation is that generally an existing expertise for same-race faces can be used for recognition of untrained other-race faces, while no such expertise exists in CP (Carbon et al., 2007).

The findings of our test battery also have some further implications for the general understanding of face perception and face processing. First, we find that better configural sensitivity relates to better face recognition ability. Koreans and Germans performed significantly better in the general face recognition task Cambridge Face Memory Test, and at the same time showed a significantly higher sensitivity to configural changes in our second test than the prosopagnosics. This importance of configural processing for holistic processing was so far only shown by disrupting configural information, e.g., by the inversion effect (Freire et al., 2000). Our finding is an important result that allows us to get further insight about which aspect of face recognition relates with being a good face recognizer. When correlating performance in the CFMT with the sensitivity to configural changes across all participants, we obtained a significant but medium proportion of shared variance of $r^2 = .24$ (which is larger than the proportion of shared variance of

$r^2 = .09$ of performance in the CFMT and sensitivity to featural changes). Until now studies looking for processes related to face recognition performance mostly correlated it to holistic processing in general (e.g., performance in the composite face task or part-whole-face-task). Different proportions of shared variance were found: either zero ($r^2 = .003$, Konar et al., 2010), or medium ($r^2 = .16$, Richler et al., 2011), or similar to our value ($r^2 = .21$, DeGutis et al., 2013). The range of results in these studies might be explained by the different measures used for face recognition (CFMT vs. own identity recognition tasks), holistic processing (composite face task vs. part-whole-face-task) and different approaches to calculate the effect scores (subtraction scores vs. regression scores, and partial vs. complete composite face design). Whether general problems in processing faces results in an inability to see subtle differences in facial configuration, whether a reduced sensitivity to configuration results in impaired face recognition ability, or whether configural sensitivity and face recognition performance are impaired by disrupting a common underlying process remains an open question. This is a decade- old, and as-of-yet unanswered issue (Barton et al., 2003) which we cannot address using our current data. Nevertheless, our results strengthen the hypothesis that configural processing is linked to face recognition ability, but the proportions of shared variance are only low to medium, which show that configural sensitivity and/or holistic processing cannot solely explain face processing abilities.

The second implication of our findings for face processing stems from the fact that we find no difference in terms of sensitivity to facial features between Koreans and prosopagnosics. This suggests that this aspect is not crucial for determining face recognition abilities. This finding is supported by the low effect size found in correlating the sensitivity to featural changes with face recognition performance (tested either using the CFMT or the face recognition performance in the object recognition task): only a small portion of the variance of face recognition abilities is explained by the sensitivity to differences in features ($r^2 = .09$ and .11 in both cases).

Overall, with our test battery we were able to replicate results of previous studies and provide new insights into the face processing disturbances caused by CP and the ORE. Thus, when a (Caucasian) prosopagnosic person tries to explain his or her condition to a (Korean) non-prosopagnosic person with the ORE ("They all look the same to you; everyone else does for me, too") this is an inexact comparison. Although the perception of Caucasian faces by Koreans and prosopagnosics observers differs, the analogy probably gives at least an idea of the problems congenital prosopagnosics (though to a stronger extent) have to face.

Acknowledgements

This research was supported by funding from the Max Planck Society, as well as from the world class university (WCU) program. We would like to thank all participants who participated in this study. Also the help of Prof. Dr. Ingo Kennerknecht in contacting the prosopagnosic participants is highly appreciated. We thank Bradley Duchaine and Ken Nakayama, Nina Gaißert, and Christoph Dahl for graciously giving us their stimulus material.

Supplementary material

The Supplementary Material for this article can be found online at: http://www.frontiersin.org/journal/10.3389/fnhum. 2014.00759/abstract

References

Avidan, G., and Behrmann, M. (2009). Functional MRI reveals compromised neu- ral integrity of the face processing network in congenital prosopagnosia. Curr. Biol. 19, 1146–1150. doi: 10.1016/j.cub.2009.04.060

Avidan, G., Hasson, U., Malach, R., and Behrmann, M. (2005). Detailed exploration of face-related processing in congenital prosopagnosia: 2. Functional neuroimaging findings. J. Cogn. Neurosci. 17, 1150–1167. doi: 10.1162/0898929054475145

Avidan, G., Tanzer, M., and Behrmann, M. (2011). Impaired holistic processing in congenital prosopagnosia. Neuropsychologia 49, 2541–2552. doi: 10.1016/j.neuropsychologia.2011.05.002

Avidan, G., Thomas, C., and Behrmann, M. (2008). "An integrative approach towards understanding the psychological and neural basis of congenital prosopagnosia," in Cortical Mechanisms of Vision, eds M. Jenkin and L. R. Harris (New York, NY: Cambridge University Press), 241–270. Available online at:

http://tdlc.ucsd.edu/research/publications/Avidan_Integrative_Approach_ 2009.pdf (Accessed June 19, 2014).

Barton, J. J. S., Cherkasova, M. V., Press, D. Z., Intriligator, J. M., and O'Connor, M. (2003). Developmental prosopagnosia: a study of three patients. Brain Cogn. 51, 12–30. doi: 10.1016/S0278-2626(02)00516-X

Behrmann, M., Avidan, G., Marotta, J. J., and Kimchi, R. (2005). Detailed exploration of face-related processing in congenital prosopagnosia: 1. Behavioral findings. J. Cogn. Neurosci. 17, 1130–1149. doi: 10.1162/0898929054475154

Bernstein, M. J., Young, S. G., and Hugenberg, K. (2007). The cross-category effect: mere social categorization is sufficient to elicit an own-group bias in face recognition. Psychol. Sci. 18, 706–712. doi: 10.1111/j.1467-9280.2007.01964.x

Blais, C., Jack, R. E., Scheepers, C., Fiset, D., and Caldara, R. (2008). Culture shapes how we look at faces. PLoS ONE 3:e3022. doi: 10.1371/journal.pone.0003022

Carbon, C.-C., Grüter, T., Weber, J. E., and Lueschow, A. (2007). Faces as objects of non-expertise: processing of thatcherised faces in congenital prosopagnosia. Perception 36, 1635–1645. doi: 10.1068/p5467

Collishaw, S. M., and Hole, G. J. (2000). Featural and configurational processes in the recognition of faces of different familiarity. Perception 29, 893–909. doi: 10.1068/p2949

DeGutis, J. M., Bentin, S., Robertson, L. C., and D'Esposito, M. (2007). Functional plasticity in ventral temporal cortex following cognitive rehabilitation of a congenital prosopagnosic. J. Cogn. Neurosci. 19, 1790–1802. doi: 10.1162/jocn.2007.19.11.1790

DeGutis, J. M., Wilmer, J., Mercado, R. J., and Cohan, S. (2013). Using regression to measure holistic face processing reveals a strong link with face recognition ability. Cognition 126, 87–100. doi: 10.1016/j.cognition.2012.09.004

Duchaine, B. C., and Nakayama, K. (2005). Dissociations of face and object recognition in developmental prosopagnosia. J. Cogn. Neurosci. 17, 249–261. doi: 10.1162/0898929053124857

Duchaine, B. C., and Nakayama, K. (2006). The Cambridge face memory test: results for neurologically intact individuals and an investigation of its validity using inverted face stimuli and prosopagnosic participants. Neuropsychologia 44, 576–585. doi: 10.1016/j.neuropsychologia.2005.07.001

Duchaine, B. C., Yovel, G., and Nakayama, K. (2007).No global processing deficit in the Navon task in 14 developmental prosopagnosics. Soc. Cogn. Affect. Neurosci. 2, 104–113. doi: 10.1093/scan/nsm003

Esins, J., Bülthoff, I., and Schultz, J. (2011). The role of featural and configural information for perceived similarity between faces. J. Vis. 11:673. doi: 10.1167/11.11.673

Fowler, D. R., Meinhardt, H., and Prusinkiewicz, P. (1992). Modeling seashells. ACMSIGGRAPH Comput. Grap. 26, 379–387. doi: 10.1145/133994.134096

Freire, A., Lee, K., and Symons, L. A. (2000). The face-inversion effect as a deficit in the encoding of configural information: direct evidence. Perception 29, 159–170. doi: 10.1068/p3012

Gaißert, N., Wallraven, C., and Bülthoff, H. H. (2010).Visual and haptic perceptual spaces show high similarity in humans. J. Vis. 10, 1–20. doi: 10.1167/10.11.2

Goffaux, V., Hault, B., Michel, C., Vuong, Q. C., and Rossion, B. (2005). The respective role of low and high spatial frequencies in supporting configural and featural processing of faces. Perception 34, 77–86. doi: 10.1068/p5370

Grüter, T., Grüter, M., and Carbon, C.-C. (2008). Neural and genetic foundations of face recognition and prosopagnosia. J. Neuropsychol. 2, 79–97. doi: 10.1348/174866407X231001

Hayward, W. G., Rhodes, G., and Schwaninger, A. (2008). An own-race advantage for components as well as configurations in face recognition. Cognition 106, 1017–1027. doi: 10.1016/j.cognition.2007.04.002

Hugenberg, K., Young, S. G., Bernstein, M. J., and Sacco, D. F. (2010). The categorization-individuation model: an integrative account of the other-race recognition deficit. Psychol. Rev. 117, 1168–1187. doi: 10.1037/a0020463

Kennerknecht, I., Ho, N. Y., and Wong, V. C. N. (2008). Prevalence of hereditary prosopagnosia (HPA) in Hong Kong Chinese population. Am. J. Med. Genet. A 146A, 2863–2870. doi: 10.1002/ajmg.a.32552

Kimchi, R., Behrmann, M., Avidan, G., and Amishav, R. (2012). Perceptual separability of featural and configural information in congenital prosopagnosia. Cogn. Neuropsychol. 29, 447–463. doi: 10.1080/02643294.2012.752723

Konar, Y., Bennett, P. J., and Sekuler, A. B. (2010). Holistic processing is not correlated with face-identification accuracy. Psychol. Sci. 21, 38–43. doi: 10.1177/0956797609356508

Kress, T., and Daum, I. (2003). Developmental prosopagnosia: a review. Behav. Neurol. 14, 109–121. doi: 10.1155/2003/520476

Le Grand, R.,Cooper, P. A.,Mondloch, C. J.,Lewis,T.L., Sagiv, N.,DeGelder, B., et al. (2006).What aspects of face processing are impaired in developmental prosopagnosia? Brain Cogn. 61, 139–158. doi: 10.1016/j.bandc.2005.11.005

Lobmaier, J. S., Bölte, J., Mast, F. W., and Dobel, C. (2010). Configural and featural processing in humans with congenital prosopagnosia. Adv. Cogn. Psychol. 6, 23–34. doi: 10.2478/v10053-008-0074-4

Macmillan, N. A., and Creelman, C. D. (2005). Detection Theory: A User's Guide. 2nd Edn. Mahwah, NJ: Lawrence Erlbaum Associates.

Maurer,D., LeGrand, R., and Mondloch, C. J. (2002). The many faces of configural processing. Trends Cogn. Sci. 6, 255–260. doi: 10.1016/S1364-6613(02)01903-4

Maurer, D., O'Craven, K. M., Le Grand, R., Mondloch, C. J., Springer, M. V., Lewis, T. L., et al. (2007). Neural correlates of processing facial identity based on features versus their spacing. Neuropsychologia 45, 1438–1451. doi: 10.1016/j.neuropsychologia.2006.11.016

McKone, E., Aimola Davies, A., Fernando, D., Aalders, R., Leung, H., Wickramariyaratne, T., et al. (2010). Asia has the global advantage: race and visual attention. Vision Res. 50, 1540–1549. doi: 10.1016/j.visres.2010. 05.010

McKone, E., Brewer, J. L., MacPherson, S., Rhodes, G., and Hayward, W. G. (2007). Familiar other-race faces show normal holistic processing and are robust to perceptual stress. Perception 36, 224–248. doi: 10.1068/p5499

McKone, E., Stokes, S., Liu, J., Cohan, S., Fiorentini, C., Pidcock, M., et al. (2012). A robust method of measuring other-race and other-ethnicity effects: the Cambridge face memory test format. PLoS ONE 7: e47956. doi: 10.1371/jour- nal.pone.0047956

Meissner, C. A., and Brigham, J. C. (2001). Thirty years of investigating the own- race bias in memory for faces: a meta-analytic review. Psychol. Pub. Policy Law 7, 3–35. doi: 10.1037/1076-8971.7.1.3

Michel, C., Rossion, B.,Han, J., Chung, C.-S., and Caldara, R. (2006). Holistic processing is finely tuned for faces of one's own race. Psychol. Sci. 17, 608–615. doi: 10.1111/j.1467-9280.2006.01752.x

Mondloch, C. J., Elms, N., Maurer, D., Rhodes, G., Hayward, W. G., Tanaka, J. W., et al. (2010). Processes underlying the cross-race effect: an investigation of holistic, featural, and relational processing of own-race versus other-race faces. Perception 39, 1065–1085. doi: 10.1068/p6608

Rhodes, G., Brake, S., Taylor, K., and Tan, S. (1989). Expertise and configural coding in face recognition. Br. J. Psychol. 80, 313–331. doi: 10.1111/j.2044-8295.1989.tb02323.x

Rhodes, G.,Hayward,W. G., andWinkler, C. (2006). Expert face coding: configural and component coding of own-race and other-race faces. Psychon. Bull. Rev. 13, 499–505. doi: 10.3758/BF03193876

Richler, J. J., Cheung, O. S., and Gauthier, I. (2011). Holistic processing predicts face recognition. Psychol. Sci. 22, 464–471. doi: 10.1177/0956797611401753

Rivolta, D., Palermo, R., Schmalzl, L., and Coltheart, M. (2011). Covert face recognition in congenital prosopagnosia: a group study. Cortex 48, 1–9. doi: 10.1016/j.cortex.2011.01.005

Rotshtein, P., Geng, J. J., Driver, J., and Dolan, R. J. (2007). Role of features and second-order spatial relations in face discrimination, face recognition, and individual face skills: behavioral and functional magnetic resonance imaging data. J. Cogn. Neurosci. 19, 1435–1452. doi: 10.1162/jocn.2007.19.9.1435

Rushton, J. P., and Jensen, A. R. (2005). Thirty years of research on race differences in cognitive ability. Psychol. Pub. Policy Law 11, 235–294. doi: 10.1037/1076-8971.11.2.235

Stollhoff, R., Jost, J., Elze, T., and Kennerknecht, I. (2011). Deficits in long-term recognition memory reveal dissociated subtypes in congenital prosopagnosia. PLoS ONE 6:e15702. doi: 10.1371/journal.pone.0015702

Towler, J., Gosling, A., Duchaine, B. C., and Eimer, M. (2012). The face- sensitive N170 component in developmental prosopagnosia. Neuropsychologia 50, 3588–3599. doi: 10.1016/j.neuropsychologia.2012.10.017

Troje, N. F., and Bülthoff, H. H. (1996). Face recognition under varying poses: the role of texture and shape. Vision Res. 36, 1761–1771. doi: 10.1016/0042- 6989(95)00230-8

Vetter, T., and Blanz, V. (1999). "A morphable model for the synthesis of 3D faces," in SIGGRAPH'99 Proceedings of the 26th annual conference on Computer graphics and interactive techniques (New York, NY: ACMPress/Addison-Wesley Publishing Co.), 187–194. doi: 10.1145/311535.311556

Wang, H., Stollhoff, R., Elze, T., Jost, J., and Kennerknecht, I. (2009). Are we all prosopagnosics for other race faces? Perception 38 ECVP Abstract Supplement, 78. doi: 10.1371/journal.pone.0003022

Yovel, G., and Duchaine, B. C. (2006). Specialized face perception mechanisms extract both part and spacing information: evidence from develop- mental prosopagnosia. J. Cogn. Neurosci. 18, 580–593. doi: 10.1162/jocn.2006. 18.4.580

Yovel, G., and Kanwisher, N. (2004). Face perception: domain specific, not process specific. Neuron 44, 889–898. doi: 10.1016/j.neuron.2004. 11.018

References

Conflict of Interest Statement: The authors declare that the research was conducted in the absence of any commercial or financial relationships that could be construed as a potential conflict of interest.

Received: 30 April 2014; accepted: 08 September 2014; published online: 29 September 2014. Citation: Esins J, Schultz J,Wallraven C and Bülthoff I (2014) Do congenital prosopagnosia and the other-race effect affect the same face recognition mechanisms? Front. Hum. Neurosci. 8:759. doi: 10.3389/fnhum.2014.00759 This article was submitted to the journal Frontiers in Human Neuroscience. Copyright © 2014 Esins, Schultz, Wallraven and Bülthoff. This is an open-access article distributed under the terms of the Creative Commons Attribution License (CC BY). The use, distribution or reproduction in other forums is permitted, provided the original author(s) or licensor are credited and that the original publication in this journal is cited, in accordance with accepted academic practice. No use, distribution or reproduction is permitted which does not comply with these terms.

118

IV. Galactose uncovers face recognition and mental images in congenital prosopagnosia: The first case report

1. Abstract

A woman in her early 40s with congenital prosopagnosia and attention deficit hyperactivity disorder observed for the first time sudden and extensive improvement of her face recognition abilities, mental imagery, and sense of navigation after galactose intake. This effect of galactose on prosopagnosia has never been reported before. Even if this effect is restricted to a subform of congenital prosopagnosia, galactose might improve the condition of other prosopagnosics. Congenital prosopagnosia, the inability to recognize other people by their face, has extensive negative impact on everyday life. It has a high prevalence of about 2.5 %. Monosaccharides are known to have a positive impact on cognitive performance. Here, we report the case of a prosopagnosic woman for whom the daily intake of 5 g of galactose resulted in a remarkable improvement of her lifelong face blindness, along with improved sense of orientation and more vivid mental imagery. All these improvements vanished after discontinuing galactose intake. The self-reported effects of galactose were wide-ranging and remarkably strong but could not be reproduced for 16 other prosopagnosics tested. Indications about heterogeneity within prosopagnosia have been reported; this could explain the difficulty to find similar effects in other prosopagnosics. Detailed analyses of the effects of galactose in prosopagnosia might give more insight into the effects of galactose on human cognition in general. Galactose is cheap and easy to obtain, therefore, a systematic test of its positive effects on other cases of congenital prosopagnosia may be warranted.

2. Introduction

Congenital prosopagnosia (CP) is the lifelong impairment in recognizing someone by their face. It is present from birth on, in contrast to acquired prosopagnosia which refers to the loss of an originally functioning face recognition system due to a brain lesion. CP is very common with a prevalence of 2.5 % in the general population [1] and thus represents the majority of prosopagnosia cases [2]. The general processes or impairments causing CP are not yet understood, but accumulating evidence suggests that CP is hereditary because it almost always runs in families [1,3,4].

3. Report of the case

In March 2008, a 40-year-old woman, LI, contacted us because she was unable and never has been able to recognize the faces of friends or even of her close family members. The self-reported diagnosis of CP (also known as face blindness) in the absence of any known events of brain damage or malformation was supported by us with a semi-structured diagnostic interview as described elsewhere [5] In addition, LI had previously been diagnosed with attention deficit hyperactivity disorder (ADHD) by a medical unit elsewhere.

In June 2011, LI had a very curious experience and contacted us a second time. On advice from a friend, she started the daily intake of one tablespoon (about 5g) of D-galactose as self-medication for her ADHD. Three days after the start of galactose intake, she began noticing clear changes in her ADHD-related symptoms. However, most surprisingly, her prosopagnosia-related symptoms improved as well; LI could suddenly recognize the faces of her family members and of persons that she had encountered repeatedly but had always failed to recognize. She could even recognize the faces of people she had not seen for many years. Furthermore, she experienced mental images for the first time in her life. LI was even somehow afraid of the vivid and colorful dreams she now had, as this was a new experience for her. Additionally, she reported an improved sense of orientation, and her need for sleep was normalized from at most 4 hours to about 8 hours per night.

A third meeting took place 3 months later, 4 weeks after she stopped her daily galactose intake following our request. At this meeting, she reported that all the effects of galactose intake had disappeared, and now described the loss of her newly acquired face recognition abilities and mental imagery, a significantly reduced need for sleep, and a decreased sense of orientation.

To assess the generality of our observation, we asked 16 further congenital prosopagnosics to take a daily dose of 5 g of galactose for 7 days and report any changes. Three of those participants had also been diagnosed with ADHD. None of the 16 participants observed any noticeable effects of galactose intake.

4.   Comment

Our participant LI showed the habitual, previously reported symptoms of CP: impaired face recognition [6], weak sense of navigation [6], and a lack of mental images, at least of faces [7]. The intake of monosaccharides is related to improvements in cognitive performance, but descriptions of their influence on visual perception especially have not been published so far [8]. Other studies have shown other positive effects of galactose on brain function. A study by Best et al [9] reported a positive effect of the supplementation of saccharides including galactose and other monosaccharides on memory performance and well-being. Further work has revealed that galactose is important for the healthy functioning of the human brain, and animal studies have shown that galactose is essential for proper myelin formation, which is necessary for functional insulation of the axons of nerve cells [10]

Galactose could play an intricate role in visual cognition. For LI, daily oral intake of 5 g of this monosaccharide had several very positive effects on her life. A possible explanation for our findings is that LI suffers from a (relative) deficiency in galactose that impairs face recognition and the generation of mental images. Supplementing galactose might compensate, a relative receptor insufficiency and alleviate her symptoms. Further, the fact that LI was able to recognize previously unrecognizable faces after taking galactose implies that familiar faces were stored in her memory but remained inaccessible before galactose supplementation (Fig. 1).

A very interesting aspect of our case is that the described effects had not occurred prior to galactose intake. Glucose, another monosaccharide and part of LI's normal diet, did not show any effects. Ritalin, which LI took to improve her ADHD symptoms before she switched to galactose, did not have the same impact either. The fact that none of the 16 other tested prosopagnosics showed similar responses to galactose as LI is in line with strong evidence that prosopagnosia is phenotypically and genetically het- erogeneous,5 which might explain why only (very) few prosopagnosics seem to benefit from galactose.

This as yet unreported effect of galactose on prosopagnosia raises many hopes. If galactose proves to be helpful for (albeit rare) subtypes of CP, it would provide a simple and cheap way to immensely improve the life of the people concerned.

Figure 1: Artist's impression of the impact of galactose on face blindness.

Ethics

The study was approved by the ethical committee of the University of Münster, Germany, protocol No 3XKenn2, amendment 2012-512-f–S. Informed consent from all participants was obtained.

References

1. Kennerknecht I, Grüter T,Welling B,Wentzek S. First report on the prevalence of non-syndromic hereditary prosopagnosia (HPA). Am J Med Genet 2006; 140A(Part A):1617–22.
2. Behrmann M, Avidan G. Congenital prosopagnosia: face-blind from birth. Trends Cogn Sci 2005;9:180–7.
3. Duchaine BC, Germine L, Nakayama K. Family resemblance: ten family members with prosopagnosia and within-class object agnosia. Cogn Neuropsychol 2007;24(4):419–30.
4. Grüter T, Grüter M, Carbon C-C. Neural and genetic foundations of face recognition and prosopagnosia. J Neuropsychol 2008;2(1):79–97.
5. Stollhoff R, Jost J, Elze T, Kennerknecht I. Deficits in long-term recognition memory reveal dissociated subtypes in congenital prosopagnosia. PLoS ONE 2011;6(1):e15702.
6. Le Grand R, Cooper PA, Mondloch CJ, Lewis TL, Sagiv N, De Gelder B. What aspects of face processing are impaired in developmental prosopagnosia? Brain Cogn 2006;61(2):139–58.
7. Tree JJ, Wilkie J. Face and object imagery in congenital proso- pagnosia: a case series. Cortex, 2010;46:1189–98.
8. Nelson ED, Ramberg JE, Best T, Sinnott RA. Neurologic effects of exogenous saccharides: a review of controlled human, animal, and in vitro studies. Nutr Neurosci 2012;15:149–62.
9. Best T, Kemps E, Bryan J. Saccharide effects on cognition and well-being in middle-aged adults: a randomized controlled trial. Dev Neuropsychol 2010;35(1):66–80.
10. Best T, Kemps E, Bryan J. Effects of saccharides on brain function and cognitive performance. Nutr Rev 2005;63(12):409–18